SOCIAL
CONSEQUENCES
OF ENGINEERING

Edited by
HAYRETTIN KARDESTUNCER
University of Connecticut

boyd & fraser publishing company
san francisco

Hayrettin Kardestuncer, editor
Social Consequences of Engineering

© 1979 by Boyd & Fraser Publishing Company. All rights reserved. No part of this work may be reproduced or used in any form or by any means—graphic, electronic, or mechanical, including photocopying, recording, taping, or information and retrieval systems—without written permission from the publisher.

Manufactured in the United States of America.

Library of Congress Cataloging in Publication Data:

Main entry under title:
Social consequences of engineering.
    Includes bibliographies and index.
    1. Technology—Social aspects.   I. Kardestuncer, Hayrettin.
T14.5.S637   301.24'3   78-23667
ISBN 0-87835-074-8
ISBN 0-87835-073-X pbk.

1 2 3 4 5 · 2 1 0 9

To Ayla

PREFACE

Engineering has always been caught between science and society. In trying to answer society's demands and in complying with scientific discoveries, it is either praised or damned. Unfortunately, both views arise out of ignorance. Not long ago, a group made up of eminent engineers and humanists got together under a Ford Foundation grant at the University of California, Los Angeles, and issued the following resolution:

> Engineering should recognize and accept a responsibility to contribute to liberal education by such means as offering an introductory course being conceptually sophisticated but not mathematically rigorous. *

Such an idea was obviously so timely, so natural and already in the minds of educators that most colleges and universities now have courses in Engineering, Technology and Society. Programs on engineering for non-engineers are becoming the most rapidly growing sector of American academe and are among the earnest institutional efforts to respond to the ever-increasing questions from society on engineering issues.

With this in mind the book has been written primarily for non-engineering students and the public who are interested in understanding serious technical problems that mankind is facing. While acquainting non-technical students with these problems, it also enlightens engineers on the restrictions imposed upon them in finding realistic solutions. Since the problems are presented in a non-mathematical way and the solutions are non-numerical, the text does not train anyone in any phase of engineering and technology but instead educates the reader on these issues.

The sequence of chapters is more or less representative of engineering developments: besides food, *shelter* has been man's first and earliest

---

*William Davenport and Daniel Rosenthal, *Engineering: Its Role and Function in Human Society*. Elmsford, N.Y.: Pergamon, 1967.

v

technological concern; *energy* and *transportation* have often been the next; *communication* and *computers* play the most important role in engineering developments; currently *pollution* has become a serious threat to our environment; while *biomedical technology* and *public safety* receive ever increasing attention, *genetic engineering* has started tinkering with life; finally man has been seriously concerned with *controlling* all of these developments *democratically*.

Although engineers, scientists, and technicians are the driving force behind technological developments, decisions concerning most of these developments are often made by those who lack understanding of technical reasoning. With the present rate of technological advancement, neither will man be able to cope with the environment nor will the environment be able to keep up with man. A basic culture and fundamental knowledge in engineering technology have never been so vital for members of all learned societies.

Considering that no engineering problem has a unique solution, there are always alternatives. One should not be surprised to find that one of the alternatives is public understanding, which may not solve the problem as the engineer would but may instead eliminate it altogether. The text is an attempt to develop such an understanding.

The text is written by engineers and scientists with facts. It not only offers an engineering culture to liberal arts students, it eventually invites them to help out engineers in finding the best alternatives to solve these mighty problems. It brings closer the liberal arts and engineering in much the same way that the fine arts and pure sciences have traditionally been brought together. Never before has the desire for mutual understanding of these two worlds become so apparent.

Many people have contributed to this text in various ways. Some reviewed the entire manuscript and offered constructive criticism while others gave helpful suggestions, ideas, and encouragement. In particular, I am thankful to Martin S. Barber (University of Colorado), William B. Berry (University of Notre Dame), Charles L. Best (Lafayette College), Philip M. Besuner (Failure Analysis Associates), Richard Brumberg (University of Pennsylvania), William P. Darby (Washington University), Carl W. Hall (Washington State University), J. Paul Hartman (Florida Technological University), Aino E. Kardestuncer (University of Connecticut), Grant Marr (Washington State University), Robert M. McKeon (Babson College), Charles Overby (Ohio State University), John Truxal (State University of New York at Stony Brook), George W.

Weaver (West Virginia University), and H. William Welch (Arizona State University).

I am also grateful to my colleagues at the University of Connecticut—Wallace Bowley, Joseph Gartner, Michael Howard, Theo Kattamis, Walter Geiger, Peter McFadden, the late Renato Nicola, Clarence Schultz, Leroy Stutzman, and Vincent Suprynowicz—for joining me in initiating and offering the course that inspired me to compose this book. Furthermore, I gratefully acknowledge the generous assistance of my former students Norman Bolle, Pat and Jay Conant, Ann Miller Brickley, and Scott Staley.

Lastly, I am indebted to my old friend Turgut I. Burakreis who, among a handful of others, foresaw yesterday the sociotechnical problems we are facing today. I have had many long and stimulating discussions with him on the topics covered in this text.

H. KARDESTUNCER

Storrs, Connecticut

CONTRIBUTORS

1 ENGINEER THE MASTER?

Hayrettin Kardestuncer
University of Connecticut

2 HABITAT

Thomas L. Saaty
The Wharton School
University of Pennsylvania

3 ENERGY AND SOCIETY

Ali B. Cambel
George Washington University

4 TRANSPORTATION AND TECHNOLOGY ASSESSMENT

Thomas T. Liao
State University of New York
at Stony Brook

5 COMMUNICATIONS

Paul Davidovits
Boston College

6 COMPUTERS

Grace C. Hertlein
California State University,
Chico
Edmund C. Berkeley
Editor, Computers and People
Berkeley Enterprises, Inc.

7 POLLUTION

P. A. Vesilind
Duke University

8 BIOMEDICAL ENGINEERING

Leslie Cromwell
California State University,
Los Angeles

9 GENETIC ENGINEERING: TINKERING WITH LIFE?

Amitai Etzioni
Columbia University and
Center for Policy Research, Inc.

10 PUBLIC RISK AND ENGINEERING SAFETY: HOW SAFE IS SAFE ENOUGH?

Alan Tetelman
University of California,
Los Angeles

11 CONTROLLING ENGINEERING DEMOCRATICALLY

Arthur Kantrowitz
AVCO Everett Research
Laboratory, Inc.

TABLE OF CONTENTS

SOCIAL CONSEQUENCES
OF ENGINEERING

Chapter One

ENGINEER THE MASTER?

1.1 INTRODUCTION

Let us begin with two anecdotes, one from the fourth century B.C. and one from our own century. Both refer to the same man: *Engineer.*

The first one belongs to Plato and it says: "Nevertheless you despise him and his art, and you will not allow your daughter to marry his son. . . . And yet, on your principle, what justice or reason is there in your refusal?" *

The second one belongs to Herbert Hoover, once the engineer president of the United States, while he was chatting with a dreadfully curious British lady: "I replied that I was an engineer. She emitted an involuntary exclamation, and said, I thought you were a gentleman!"†

Much has been said about engineers and engineering in every epoch. As in the two anecdotes cited here, in most cases, "engineer" and "engineering" are interchanged freely. Engineering is personalized hypothetically and referred to as "engineer." It would be appropriate if we could give a definition of "engineer" and of his profession "engineering," Definitions are, however, often misleading. No matter how broad it may be, a simple definition cannot encompass all phases of engineering. The

---

*Friedrich Klemm, *A History of Western Technology:* Cambridge, Mass.: MIT Press, 1964.
†Herbert Hoover, *Years of Adventure:* New York: Macmillan, 1951.

1

following definition, however, has been adopted by the Engineer's Council for Professional Development: Engineering is the profession in which a knowledge of mathematical and natural sciences gained by study, experience and practice is applied with judgment to develop ways to benefit mankind.

Considering that economy and beauty often adversely affect each other, many great engineers who sought beauty throughout their lives would find such a definition incomplete.

Another famous definition of engineering, "the art of directing the great sources of power in nature for the use and convenience of man . . ." has been given by Thomas Tredgold, a British architect (1788–1829), and is contained in the charter of the Institute of Civil Engineers. This definition, however, seems to exclude all military engineering including some works of Leonardo da Vinci. Here we shall not try to define engineer and engineering but try to develop some understanding of them.

The engineer, in general, is a person who puts to work what scientists discover, technologists suggest, and the public demands. He used to be an elite and educated individual. Not so long ago, names like Roebling, Eads, Brunel, Amman, and Steinman* were carved on the portals of the bridges they designed. They were the individuals who met with elected officials of a city, raised funds for the project, guided the design, and supervised the construction. Today, much larger bridges, e.g., the Verrazano Bridge linking Brooklyn and Staten Island and the Bosporus Bridge between Europe and Asia, are built yet no single individual truly deserves the engineering honor such as was bestowed upon John Roebling (12) for the great Brooklyn Bridge (1889).

For a medium-sized bridge, for instance, there are a number of engineers working together. For example, the *foundation engineer* deals with soil and rock mechanics and piles; the *substructure engineer* designs piers, abutments,and footings; the *superstructure engineer* designs floor beams, stringers, and girders; the *highway engineer* deals with pavements, bituminous surface, and drainage; the *electrical engineer* lays out the electrical conduits and illuminations. Additionally, there is a *project engineer,* an *architect,* and a number of *draftsmen.* None of them claims to be the sole designer of the bridge. Rather, each assumes the responsibility for the portion of the bridge he designed. When all the pieces are put together, the result of such a group effort may become an engineering feat: a sus-

---

*Noted bridge engineers of 1850–1950.

pension bridge, a skyscraper, an oil refinery, a nuclear power plant, an air transport, a space vehicle, a supertanker, and so on.

Who puts the pieces together? Not a single individual but a team: the project engineer, chief engineers, managers, etc. The function of the engineer, then, is designing small components and transferring information from one component to another. The questions of why and how an engineering project comes into existence, however, remain to be answered.

If engineering is the leading profession in the conquest and utilization of matter, energy, time and space, the engineer himself is not the leader. He is the grinding but not the deciding force behind this conquest. The decisions on most important engineering projects are sometimes made by those who are not sufficiently knowledgeable in engineering. This apparently has been felt in the past as well. According to Professor Hardy Cross (an eminent civil engineer of the forties), for example, "There are groups of self-styled engineers who are telling the country how valuable they are and how accurate are their conclusions. These men attempt, often consciously though sometimes unconsciously, to give the impression that they deal with measurable data from which definite laws useful to mankind may be deduced. They often call this leadership." One may wonder why in recent years so much has been said against engineers and technologists. Environmentalists call them corporate mercenaries, intellectuals sneer at them as robots devoid of feeling, humanists attack them in much the same way that pacifists condemn wars.

Contrary to general belief, some technological inventions are the work of non-engineers, and many patent holders (including some very significant ones) do not belong to the engineering profession. Charles Kettering, vice president of General Motors in the forties, once said, "Some years ago a survey was made in which it was shown that, if a person had an engineering or scientific education, the probability of his making an invention was only about half as great as if he did not have that specialized training. . . . An inventor is a fellow who doesn't take his education too seriously."

If there is any truth in this, the following might be some of the reasons for it. First, engineering courses are well structured with facts and numbers, discourage unorthodox thinking, and allow no risk. An engineering student is continuously trained to use a factor of safety in all calculations, principles, and methodologies.

The second reason is his work. Most engineers in industry are asked to do the same (or similar) work over and over. This develops proficiency and is good for the company because the engineer becomes an expert. However, after a certain point he becomes bored with the repetition. Although he may be proficient, he is no longer efficient. Working in the same field continuously tends to stifle creativity. Once you have done it, it is difficult to do it over from a completely new perspective.

If a single experiment is worth a hundred theories, it seems very logical to ask a young engineer to perform calculations according to an employer's proven methodologies. When a person has been trained to comply with very rigid rules or physical laws, it is understandable that he would give no argument on issues contradictory to his artisanship. While this may seem to the psychiatrists to be an unhealthy rigidity, it is admirable persistence, self-discipline, and professional devotion to the engineers.

If there is a "weak spot" in engineering, it may perhaps be attributed to education. It has been recommended that engineering education should be upgraded by designing a pre-engineering undergraduate curriculum prior to professional school, as is done for lawyers and doctors. This may create many other problems. Besides that, it has been claimed that to alter engineering curricula is harder than to move a cemetery in New England.

The value of liberal arts studies in engineering education, however, has long been acknowledged. Most engineering colleges today encourage their students to choose at least one course each semester from the liberal arts. Charles Susskind, for instance, refers to engineering education as "a more liberal, that is a broader education than most curricula in the liberal arts" (19). While most educators agree that courses in history, philosophy, art, literature, business, politics., etc. allow engineering students to think in terms of alternatives and to understand the world we live in, some engineers question their value.

Myron Tribus, the director of MIT's Center of Advanced Engineering Studies, describes three different worlds where engineers might possibly live: *-ics, -ing,* and *-tion* worlds.* The first one is the world of academics where the engineer began his studies on mathematics, physics, mechanics, electronics, etc. He then enters the second world, a world that is basically concerned with doings: planning, analyzing, designing, testing, manufacturing, etc. The last world is the world of politi-

---

*Address presented at the Pennsylvania State University gathering, Technology and Society on the Campus, October 1975.

cians, which is characterized by dealing with the institutions and functions of the society. This is the world where communications, habitation, wealth, health, and energy distributions, nutrition, education, and so on, are handled. He claims that engineering professors seldom get beyond the -*ics* world and engineers beyond the -*ing* world. Yet those in the -*ics* world are largely illiterate on issues of the other two worlds. He also claims that only 1.4 percent of the foreign service officers in the State Department have an education in one of the physical sciences including engineering. Considering that some of the issues that the State Department is occupied with are: nuclear weapons, communications, technology transfer, food, pollution, transportation, habitation, the environment, power development, etc., the gap between these three worlds becomes apparent.

According to this classification, the engineers are caught between the world of academicians and the world of politicians. In the case of a grave situation, the academicians blame engineers for faulty interpretation of scientific facts and the politicians accuse them of misunderstanding what they have said. Since most engineering products are the result of scientific discoveries and are the response to society's needs, those in the -*ics* world as well as those in the -*tion* world have always claimed that they know and understand engineering. When one discovers that only a handful of academicians and politicians have ever been exposed to engineering training, it is hard to justify their criticism.

1.2 GREAT SUCCESSES AND MIGHTY FAILURES

Behind most engineering accomplishments there has been some failure, enormous struggle, and endless agony. Some of these accomplishments took generations to build and some took thousands of lives. To hold millions of gallons of water behind a wall, to make a hole through the mountains, to cross a bridge from Europe to Asia, or to land a man on the moon were not the results of ingenious ideas or overnight inventions. Successes in engineering do not take place in a laboratory as they do in most sciences or on paper as in arts and literature. Engineers always live with the fear that they might have overlooked the possible failure of the least likely component of the system. On January 27, 1967, three Apollo astronauts perished in a launch-pad fire caused by a little electric arc in the wiring. Such an arc would cause hardly any damage under ordinary circumstances but in this case the cabin atmosphere was intensified with pure oxygen which is very flammable.

Another nightmare took place in 1971. The three Russian *Soyuz 11* cosmonauts died during the re-entry when an apparently imperfect command module seal allowed a rapid drop in cabin pressure. There is no doubt that the engineer's laboratory is the real world. His experiments take place out in the open where all can see them.

In his memoirs, Herbert Hoover, engineer president of the United States, compares engineers with other professionals:

> His acts, step by step, are in hard substance. He cannot bury his mistakes in the grave like the doctors. He cannot argue them into thin air or blame the judge like the lawyers. He cannot, like the architects, cover his failures with trees and vines. He cannot, like the politicians, screen his shortcomings by blaming his opponents and hope that the people will forget. The engineer simply cannot deny that he did it. If his works do not work, he is damned.

Food and shelter have at all times been man's most fundamental needs. Most probably, then, the first engineers were irrigators, architects, and the tool makers for farming, hunting, and fighting. Although the professional names were attached much later, these early engineers can be classified as *civil* and *military*.* While the civil engineers were busy elevating man's comfort, the military engineers had to find ways to defend it. No doubt the struggle between the "have" and the "have not" will continue forever. Man's desire for conquest will give birth to ever newer branches of engineering in the years to come. Some of the newest ones entering the profession are biomedical engineering and genetic engineering.† While the former does seem to serve man's comfort, it is simply too early to say anything about the latter.

If one agrees that civilization and engineering grew together, then the "cradle of civilization" may also be considered the birthplace of engineering. That place is called Mesopotamia, the land between the two rivers, Tigris and Euphrates. Evidently Mesopotamia was in the middle of the golden belt of ancient engineering, which stretched from the Mediterranean through Egypt, Asia Minor, and India up to Southeast

---

*The first use of the word "civil" to differentiate from the "military" belongs to John Smeaton, 1750, in England.

†The reader can refer to chapters 8 and 9 for biomedical and genetic engineering respectively.

Asia. In fact, the original seven wonders of the world, as listed by Antip-atros of Sidon (300 B.C.)—the pyramids of Egypt, the Hanging Gardens of Babylon, the statue of Zeus by Phidias at Olympia, the temple of Artemis at Ephesus, the tomb of King Mausolus of Caria at Halicarnassus, the Colossus of Rhodes, and the Pharos of Alexandria—were all in the Middle East (18). With the exception of the pyramids, none of these wonders remains today.

The first recorded engineering work of early Egypt was the wall of the city of Memphis, the capital of the Old Kingdom on the western bank of the Nile delta. The largest pyramid (King Khufu's pyramid) measures 756 feet square at the base and rises to a height of about 480 feet. The cathedrals of Florence and Milan, St. Peter's of Rome, St. Paul's in Lon-don, and Westminster Abbey could all be placed in it (18). It is made of approximately 2,300,000 blocks of stone each weighing 2.5 tons. Except for the Great Wall of China, it is the largest manmade structure. The base is almost a perfect square, with a seven-inch tolerance. It is also oriented within less than one-tenth of a degree of the true north-south and east-west directions. Such accuracy can hardly be accomplished even today.

There is no need to recite here all the great engineering accomplish-ments. Most of them are around us. We live with them and are in daily contact with them: dams, buildings, bridges, communication devices and computers, all forms of transportation including space travel, medical instruments, and so on. Undoubtedly there are many more to come. At the annual meeting of the American Academy for the Advancement of Science in Washington, D.C., February 13–17, 1978, a variety of futur-istic engineering projects were discussed. The majority of the six thou-sand participants were particularly concerned about "macro-engineering projects," the kind of project so large that making one involves a con-siderable slice of the gross national or international resources. For in-stance, one project discussed was the orbiting solar station, which would consist of a three-by-ten-mile array of solar cells to convert sunlight into electricity and beam it to earth in microwave form. A receiver on the ground, six miles in diameter, would convert the microwaves into elec-tricity. It is estimated that by the middle of the twenty-first century 112 such satellites could provide about half of the world's electricity require-ments. The cost of an orbiting station (excluding development costs) would be ten billion dollars.

Among the many other projects the most notable was a twenty-one-minute underground ride between New York and Los Angeles for about six dollars. In this project, which was called "planetran," the trains are magnetically suspended in a vacuum and are powered by electromagnetic impulses. In view of the fact that the present Interstate Highway system has already cost $400 billion, the proposed planetran with a projected cost of $250–500 billion was not considered impossible.

There have been many failures behind most engineering accomplishments. Earlier ones have already been forgotten. We admire only the newest accomplishments or those that have lasted until our times. For example, it is recorded that the magnificent dome of St. Sofia (537 A.D.) in Istanbul collapsed twice before completion. So did the famous Quebec Bridge over the St. Lawrence River, which was the largest cantilever bridge in the world (1800-foot main span, including a central suspended span of 675 feet; altogether 3529 feet long and 150 feet wide). The construction started in 1899 and on August 29, 1907, the bridge collapsed during the construction. Seventeen thousand tons of steel and eighty-six workers went down. The second failure took place in September 1916 while the center span was being lifted into place. The bridge finally opened to traffic in November 1917. The Quebec Bridge today stands as an example of engineers' courage, determination, and persistency to conquer the forces of nature.

Another example of bridge failure is the Tacoma Narrows Bridge in Washington in 1940. This handsome suspension bridge was modeled after the Bronx–Whitestone bridge but was longer in span (2800 feet between the towers as opposed to 2300 feet); it was ruptured by wind four months after completion. The failure was due to excessive vertical and torsional oscillations caused by very narrow stiffness girders. The failure was a shock and a challenge to American engineers. Yet the disaster taught engineers to understand better the mysterious phenomenon of aerodynamic instability. A few years later and with the knowledge gained from the initial failure, a new bridge was built at the same location.

The mightiest failures belong to dams. On May 31, 1889, South Fork Dam near Johnstown, Pennsylvania, collapsed with 0.5 billion cubic feet of water behind it. The failure was due to engineering error. The capacity of the spillway was not large enough to accommodate an unusually heavy rainstorm. The reservoir was emptied in forty-five minutes. It took over 2,200 lives, making it the worst dam failure in American history.

In European history, one of the biggest dam disasters belongs to the Vaiant Dam on the Piave River in Belluno, Italy, on October 10, 1963. It was the world's highest arch dam (873 feet high and 72 feet thick) and contained 50,000 tons of concrete. The dead numbered in the thousands.

The most recent dam failure in the United States took place in Idaho on

June 5, 1976, when the Great Teton dam collapsed. Eighty billion gallons of water rushing like ten-foot tidal waves took the lives of eleven people, ripped topsoil from one hundred thousand acres of fertile farmland, drowned thirteen thousand head of cattle, and destroyed thousands of homes. In all, it caused a billion dollars' worth of damage. According to a report submitted on December 31, 1976, to the Department of Interior by the nation's outstanding soil and dam engineers, the dam failed because of internal erosion. The design of the dam did not adequately take into account the foundation conditions and the characteristics of the soil.

As far as the failure of dams is concerned, a careful analysis of more than three hundred dams from all over the world indicates that 35 percent of failures are caused by floods in excess of the spillway design, 25 percent by foundation problems, and the remaining 40 percent by various causes, e.g., faulty design and construction, improper operation and maintenance, use of inferior materials, etc.*

Sometimes an engineering feat might turn out to be a catastrophe in another sense. For instance, the Aswan Dam in Egypt, which was put into operation in 1971, is the largest in the world and one of modern engineering's grandest accomplishments. Despite the dam's apparent benefits, the silting up behind the dam robs the water that flows into lower Egypt of essential nutrients. Furthermore, health authorities note that schistosomiasis, which is spread by snails in the water and is the most prevalent disease in Africa, has increased drastically.

In April 1973, environmentalists published "Disasters in Water Development," an assessment of thirteen public projects in the United States that they considered among the most environmentally destructive to be built with federal money. Strangely, the recently failed Teton Dam was among those mentioned. The report argues that the dams and channelizations or water diversions would flood wilderness, damage water quality, inundate priceless archeological remains, and harm wildlife.

Another example of engineering feat turned failure is the Avery Fisher Hall in Lincoln Center. The acoustics of this magnificent structure had been a serious problem since the hall was built in 1962. Fourteen years later, the interior of the hall was completely rebuilt at a cost of $6.4 million, which is more than half of the original cost of the entire hall.

All of these failures, and many more that are not cited here, have paved the road of today's engineering successes. Without them and the engineer's strong desire to rebuild with lessons learned from the failures, we

---

*D. H. Manning, *Disaster Technology:* Elmsford, N.Y.: Pergamon Press, 1976.

would not be at the level of engineering technology that we are today. So, in the words of Hegel, "We may affirm absolutely that nothing great in the world has been accomplished without passion" and risk.

1.3 MISUSE OF ENGINEERING?

Like most other professions, engineering may also be misused, mispracticed, and mismanaged. If its proper use is the one that benefits mankind, then all other uses of it can be considered misuse. Some misuses might be intentional (e.g., the case of war technology); some might even be in the form of over-use (e.g., driving a 260-horsepower car to go buy an ice-cream cone). Considering that the word "engineer" is derived from the Latin *ingeniator*, which in turn implies one who invents *ingenia* or war machines, the origin of engineering was not at all intended for the benefit of mankind. Devices were designed and built for the purposes of hunting, attacking, or defending. Archimedes did not use sun rays to heat residential houses or public baths but to destroy the enemy vessels. He also invented catapults, which armies used for several centuries to hurl great rocks at their enemies. Even after the invention of gunpowder, engineers like Leonardo da Vinci and Agostino Ramelli designed and built crossbow-type catapults and dart throwers for the same purpose. The magnificent fortresses we admire as engineering feats were not built to house their inhabitants but to protect them from neighbors. The only manmade structure that can be seen from the moon, the Great Wall of China, was built merely to stop the invaders from Mongolia; and the famous Lighthouse of Alexandria, one of the original Seven Wonders of the World, was not built to watch after the seamen stranded at sea but to watch for approaching enemy fleets. Even the first atomic bomb that exploded in Hiroshima on August 6, 1945, and took eighty thousand lives, is among the greatest engineering accomplishments of this epoch, yet its benefit to mankind is still debatable. The philosophy expressed by President Eisenhower in his farewell address in January 1961, that "our arms must be mighty, ready for instant action, so that no potential aggressor may be tempted to risk his own destruction," has been shared by leaders of the world at all times. During the fiscal year 1974, for instance, only $4,800 million out of $16,800 million in research and development was spent for civilian needs, the rest having gone to military research conducted by the Department of Defense. An even larger share of money and resources is allocated to the same purpose by other nations. In March,

1977, President Carter of the United States made a very comprehensive "arms limitation" proposal to the Soviet Union. The proposal basically consisted of the freezing of development of new weapons, reducing strategic nuclear weapons, and controlling war technologies. Since the proposal meant too much sacrifice on the part of the Soviet Union (their war technology apparently is on a much larger scale than the United States'), it was turned down immediately.

In addition to military use, engineering technologies geared to producing certain drugs, liquors, tobacco, etc. are also questionable regarding their benefit to man. Similarly, some pesticides such as the well-known DDT have been found to possess more harmful side effects to our environment than their intended benefit. The same thing might also be said for some food additives and synthetics.

In deciding the proper use or misuse of a particular engineering technology, one should study its, *harm* versus its *benefit* projected in time, space, and cost. Some engineering products might be very beneficial now but harmful tomorrow (such as the Aswan Dam) or they might be very beneficial at certain quantities and very detrimental thereafter (such as speed and number of motor vehicles). This would indicate that an engineering accomplishment in one sense might very well be a failure in another sense. It is not necessary to mention all the great engineering accomplishments of the past which were, or were intended to be, used by tyrants against mankind; they are simply too numerous.

The writing over the portals of the 1893 Chicago World's Fair—"Science Explores, Technology Executes, Man Conforms"—may not necessarily be true. Society exercises more demands than conformity. The engineer, for instance, is perfectly capable of designing and building much more economical, safer, smaller, and slower cars, but would man conform by buying them? The same engineer can design small villages (communities) where houses have common walls and roofs (the best form of energy-saving and social living) instead of scattered dwellings for the sake of privacy.* Would man be attracted to them? How can man accept mass transportation with trains and buses that are safer and more economical and beneficial to our environment, instead of individual motor vehicles for the sake of absolute freedom of movement?

Until recently, man was granted boundless freedom in consuming the world's natural resources and discharging wastes. In the opinion of some, "The fact of the matter is that all minerals in the earth's crust exist in

---

*Refer to chapter 2 for the concept of Compact City.

virtually inexhaustible supply. If you add them all up, even the scarcest of them, such as gold, you find that they exist in the earth's crust in enormous quantities, far beyond our imagination to consume" (5). No matter how absurd, this philosophy has been shared by many. On March 27, 1977, however, President Carter urged the nation to great sacrifices on energy, a "moral equivalent of war," to bar a national catastrophe. It was obvious in his address that some of our engineering accomplishments have been misused (overused).

The *Encyclopaedia Britannica* refers to the engineering profession by saying, "The engineer's principal work is to discover and conserve natural resources of materials and forces, including the human, and to create means for utilizing these resources with minimal cost and waste and with maximum useful results." If this is true, then why do most industrial societies waste more materials and men than the primitive societies? Is this another example of misuse of engineering?

In recent years, many claim that while industry has been rushing forward with "new accomplishments," the planet Earth has been totally neglected. In November of 1956, an eight-mile stretch of concrete roadway was opened to traffic near Topeka, Kansas, marking the beginning of the Interstate Highway system. The system, which is totally funded by the Federal Highway Aid Act of 1956, is now near completion (38,000 out of 42,000 miles are open to traffic).

The impact of the system on national life and life styles has been immeasurable. As Walt Whitman said, "I inhale great draughts of space, the east and west are mine, and the north and the south are mine." While saving tens of thousands of lives (dropping from 6.28 deaths in 1956 to 1.55 in 1974 for every hundred million miles traveled), the system has nevertheless been strongly criticized for ravaging nature and destroying our cities as well as for nourishing the cancerous development of suburbs and exurbs. The pressure from environmentalists in the late sixties and seventies to "stop latticing the face of the earth with concrete" has held back some of its projects because of their environmental unsoundness. For example, the Lower Manhattan Expressway, the interstate connections inside the Washington Beltway, certain portions of the project in San Francisco, and I-210 outside New Orleans have all been stopped. There are many questions remaining to be resolved before deciding whether "the cloverleaf will become our national flower."

This is the first generation to realize that technological progress can not continue as it has in the most recent past. Society did not act on pollution and enviroment until the London Smog killed four thousand people in

1952, and we had to wait until the year 1969 to come up with a Public Law 91-190, the Environmental Policy Act, requiring individual firms and public agencies to prepare environmental impact reports on their projects. Furthermore, in 1972, a congressional Office of Technological Assessment (OTA) was established to protect man against his own inventions.

At a conference held in Washington, D.C., in March 1977, scientists agreed that genetic engineering is closer to reality than most people realize, but they disagreed sharply on whether its advent would be more of a blessing or a disaster. In spite of strong opposition, research in genetic engineering toward the insertion of new genetic material or the replacement of certain genes for making fundamental changes in a person's biological or perhaps behavioral characteristics undoubtedly will be carried on. They also agreed that genetic engineering, which deals with recombinant DNA (deoxyribonucleic acid, which is the active substance of the genes of all living things and thus governs the heredity of all life), will be the most important social issue of the next decade. It seems too early to say whether the current research and its outcome will, tomorrow, be considered one of the misuses of engineering. *

The blame for misuse of engineering seems often to be shifted to the engineers. Eric Sevareid, in his article "Slide Rule Wizards Need Re-education," † accuses engineers even of "bricking up office windows because the slide rule figures prove that air-conditioning works more efficiently in the absence of glass" (3). Yes, it is true that air-conditioning units work more efficiently in the absence of windows, but it is not engineers who make the decision on whether buildings should have windows or not. It is the public that is sold on one idea or the other. ‡

1.4 NATIONAL, INTERNATIONAL, AND UNIVERSAL BOUNDARIES OF ENGINEERING

The outcome of most engineering projects is no longer confined to particular geographical locations where the project actually takes place. The majority of developments are being felt in far-away places as though the world were getting smaller and smaller.

Man first possessed and exercised his rights on a piece of land. Then

---

*For further information on genetic engineering, refer to chapter 9.

† More of this topic is treated in section 1.5.

‡ Refer to chapter 2 for the concept of Compact City.

the shorelines were claimed by nations and in some countries even by individuals. Nations agreed, argued, and fought over offshore limits. Some nations have already extended their zone of sovereignty to two hundred miles off the coast line. The oceans which for centuries were regarded as "belonging to no one" are now about to be divided. It will not be very long before we parcel the moon and the nearest planet as well.

While these boundaries physically extend farther and farther, their actual definitions get more and more ambiguous. Yesterday any nation could experiment on any engineering project on its own territory. Today we must not only ask the permission of our immediate neighbors but on some issues we should also consult the rest of the world. Yesterday, for instance, any nation could dam its rivers. Today the Aswan Dam could not have been built without the permission of Sudan. Even as early as 1928 in building Jebel Aulija Dam in the Sudan, Egypt and Sudan had to agree to give control of the Nile waters to Egypt. Similarly, the dispute between Iraq and Syria over the Euphrates, the "shared river," has always been a problem.

Although the examples cited above basically interest (geographically at least) only two nations, there are other projects in which many nations have the right to interfere. Pollution, for instance, is one issue. To this effect, in December 1976, the members of the International Rhine Commission (France, Luxembourg, the Netherlands, Switzerland, and West Germany) signed their first binding agreement.*

It seems to be acceptable to all that no proprietor, whether an individual, a state, or a nation, should have the right to pollute the "commons," nor even their own proper land, air, and open waters. The state of Illinois, for example, filed a suit in 1972 against the city of Milwaukee for dumping raw and partly treated sewage into Lake Michigan. As is clear from this example, pollution is no longer simply "fouling one's own nest."

Another issue that concerns almost everyone is nuclear testing. It is claimed that radiation fallout was felt as far away as the United States from one of the nuclear tests of 1976 in mainland China. One should

---

*It has been noted that when Julius Caesar's armies pushed to the west bank of Europe's largest river in 51 B.C., local tribes told them it was called "Rhenos," a word that may have meant "clear." The Rhine salmon, a delicacy in the nineteenth century, is now extinct and the river itself came close to dying in June 1969 (*New York Times,* March 13, 1977).

never forget that in case of heavy doses of nuclear radiation, even space is no escape. Considering that the earth has measurable dimensions (diameter, perimeter, etc.), it certainly is conceivable that its resources are also measurable (limited). The most recent concept, "inexhaustible resources of the oceans," is no longer a valid argument. If it were, then what would happen if engineering technology were to help poor countries reach the level of the United States, where one-half of all the world's resources are consumed by only six percent of the world's population? In the words of Nobel laureate Glenn T. Seaborg, "Today we must think of the New World in terms of the entire world, as a community of mankind where the future lies in pursuing the belief that knowledge—universally obtained, widely shared, and widely applied—is the key to the viability of the human race and the earth that supports it." *
What we would like to see, however, is that engineers not be asked nor be tempted to conquer the planet earth as they were in the past, but to try to make it a home where we can all live.

1.5 CONSENT OF THE GOVERNED

One of the most profound definitions of democracy is contained in the words "consent of the governed." In order for consent or dissent to exist, there must be an issue. The issues that we are concerned with in this text are, of course, the technological issues. Considering that in all our major technological developments, the consent of the public is always sought, it would seem to be very appropriate for us to look deeper into the word "consent" and try to understand what it really means. We are certainly not interested in its literal meaning, which is "agreeing in sentiment or opinion"; we are, rather, interested in how consent is attained.

First of all, consent on an issue can be given by those who are (a) uninformed, (b) misinformed, or (c) reasonably well informed. The question then arises as to whether all three consents should be weighed in the same manner. Although in democratic societies, on any vital issue, e.g., nuclear research, the consent of a farmer and of a nuclear physicist are equally weighed, the reasons behind their consent most often have no relation whatsoever; each gives his consent for different reasons. Con-

---

*G. T. Seaborg, "Science, Technology and Development: A New World Outlook," *Science,* vol. 181, July 6, 1973, pp. 13–19.

sidering that one's reasons could be more "reasonable" (if there is such a thing as absolute truth) than the other's, it would then seem to be "most reasonable" to govern with the consent of those who were supposedly more knowledgeable on that issue. This is unacceptable in democratic societies. Even though we know that democracy is not the best form of government, man has not yet been able to come up with a better one. Until he does, we must accept the fact that every adult person has equal say on all issues regardless of how scientific these issues might be or how ignorant he or she is regarding them.

Of course, on no issue can all men have the same degree of knowledge. Therefore, the consent can never be uniform (qualitatively) nor unanimous (quantitatively). Dissent, then, will always have its place in our society. It is the "yardstick" of democracy by which we can measure whether the consent is meaningful or not.

In engineering, facts are most often proven to be correct by starting with an assumption to the contrary. Failure is understood and studied prior to design of a safer structure. Otherwise, how safe is safe enough would be a much bigger question than it presently is.*

If consent on an engineering issue is truly based on misinformation or ignorance, would not the outcome of it be unbeneficial or detrimental to society? If so, would not such a decision then be contradictory to the definition of the engineering profession, which is "to benefit mankind," etc.? No one would disagree with the idea that the public must be well informed on all technological issues so that its consent (or dissent) will be meaningful. Such an idea creates bigger problems than the issue itself; how much knowledge is enough knowledge? How do we determine a useful level of such information without killing it? If one can not draw a definite line to it, can we then say that "some" is better than "none"? Or is it true what Max Weber† said, "Every bureaucracy seeks to increase the superiority of the professionally informed by keeping their knowledge and interest secret"?

In addition to "some" or "none," from time to time we also witness the fact that not only the governed but those who govern are misinformed or kept in darkness. In spite of being so physical and factual, even engineering often allows room for different opinions on real issues. There comes the ambiguity of the definition of intentional or unintentional misinformation or even of the "right" and the "wrong" informa-

---

*The reader is advised to refer to chapter 10 for the safety of engineering systems.

†German economist and sociologist (1864–1920).

tion. Such an ambiguity permits the creation of a "magic box" in our system. There the "vested interests" are hidden and nourished (refer to Figure 1.1). It is most probable that in the box there exists "a tiny enclosed fraternity of privileged men elected by no one . . . [who] wield a free hand, selecting, presenting and interpreting the great issues of our nation."* Those who remember or who have reviewed the political

FIGURE 1.1

FLOW CHART OF TECHNOLOGICAL DEVELOPMENTS
IN DEMOCRATIC SOCIETIES

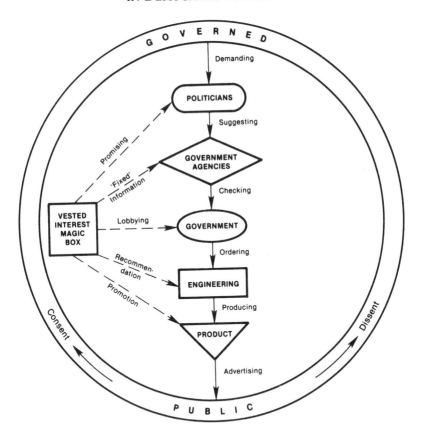

---

*The words of Spiro Agnew, former vice president in the Nixon administration, quoted by Barry M. Casper, "Technology Policy and Democracy," *Science.* vol. 194, October 1, 1976.

events of the seventies may not have any doubt that these words of Agnew, after all, carry some truth.

In totalitarian systems, such a magic box is almost wiped out, but with greater expense than benefit. Such a magic box is vital for democratic systems. There the vested interests, whether "good" or "bad," are hidden. It is the steering mechanism for our technological developments; its presence is more "natural" than the system itself. There free enterprise plays its game; there the challenge, the competition between opposing ideas, is met. It is, of course, feasible that sometimes the public might be sold on the wrong idea. However, this should not be a great concern nor one to alter democratic principle in our technological developments. An old proverb states that "those who love roses do not mind the thorns."

One can see from Figure 1.1 that this magic box plays a role at every level of democratic systems. Not only are engineering developments controlled by it, even the data furnished to the government agencies about engineering products passes through it. Since the government agencies are not capable of doing research on and experimenting on a particular issue as well as the industry itself is, they have no other alternative but to cast their judgment based on the data furnished to them by the very industries involved. This indicates to us that "the question of policy" and "the question of technology" can hardly be separated from each other. If this is so, we then believe that no expert should have a greater voice in the resolution of technological questions than the ordinary citizen or layman. This may even result in selling (for a reasonably short period of time) the wrong product to the public. Eventually, however, consent is replaced by dissent. One thing is sure: as Hegel worded it, "The idea of an epoch always finds its appropriate and adequate form."

On March 25, 1977, President Carter invited twenty ordinary citizens (housewives, construction workers, truck drivers, bank clerks, etc.) to meet and discuss the nation's energy problems with the "energy policy makers" of the administration. Such an action seems to go hand in hand with the idea expressed in the following statement of Abraham Lincoln: "With public sentiment, nothing can fail. Without it, nothing can succeed." * This text is a modest attempt to develop such a sentiment on engineering issues.

---

*R. W. Lee, *Politics and the Press:* Washington, D.C.: Acropolis, 1970; p. 56.

1.6 EXERCISES

1. Name a few great ancient engineers and cite some of their work.

2. Name a few great engineers of this country and cite some of their work.

3. Write a short essay on the interaction between engineering and the liberal arts in your institution. To what extent are non-engineering students exposed to engineering topics in general? Express your recommendations along with your findings.

4. Name an engineering failure (not mentioned in the text) and identify the reasons, with factual material.

5. Name an engineering feat that is also claimed to be a failure. Explain.

6. Name a few examples in which you think engineering technology has been misused. What do you think the remedies should be?

7. How to you think the "great sacrifices" to which President Carter urges the nation can be accomplished?

8. Argue either for or against our present Interstate Highway system.

9. Name an engineering project that has been a matter of concern for at least two nations or states. What has been the final decision?

10. Name some international organizations that have a strong voice in technological disputes between nations. Cite briefly a few examples.

11. In democratic societies, how is "consent" on major engineering developments attained?

12. Express your opinion (either pro or con) on the necessity of a magic box (as in Figure 1.1) for technological developments.

1.7 BIBLIOGRAPHY

1. Armytage, W. H. G. *A Social History of Engineering.* Cambridge, Mass.: MIT Press, 1961.
2. Davenport, William, and Daniel Rosenthal. *Engineering: Its Role and Function in Human Society.* Elmsford, N.Y.: Pergamon Press, 1967.
3. Dorf, Richard C. *Technology, Society and Man.* San Francisco: Boyd & Fraser, 1974.
4. *Engineering for the Benefit of Mankind.* National Academy of Engineering, 1970.
5. Flaxman, Edward. *Great Feats of Modern Engineering.* Arno, N.Y.: Books for Libraries Press, 1967.
6. Florman, Samuel C. *The Existential Pleasures of Engineering.* New York: St. Martin's Press, 1976.
7. Gendron, Bernard. *Technology and the Human Condition.* New York: St. Martin's Press, 1972.
8. Hamilton, David. *Technology, Man and Environment.* New York: Charles Scribner's Sons, 1973.
9. Hammond, Rolt. *Engineering Structural Failures.* Odhams Press, 1956.
10. Kingery, R. A., R. D. Berg, and E. H. Schillinger. *Men and Ideas in Engineering.* Urbana: University of Illinois Press, 1967.
11. Layton, Edwin T., Jr. *The Revolt of the Engineer.* Cleveland: Press of Case Western Reserve University, 1971.
12. McCullough, David. *The Great Bridge.* New York: Simon & Schuster, 1972.
13. Merrit, Raymond H. *Engineering in American Society.* Lexington: University Press of Kentucky, 1969.
14. Metz, L. Daniel, and Richard E. Klein. *Man and the Technological Society.* Englewood Cliffs, N.J.: Prentice-Hall, 1973.
15. Morgan, Arthur E. *Dams and Other Disasters.* Boston: Porter Sargent, 1971.
16. Ritti, Richard. *The Engineer in the Industrial Corporation.* New York: Columbia University Press, 1970.
17. Roadstrum, W. H. *Excellence in Engineering.* New York: John Wiley & Sons, 1967.
18. Sprague de Camp, L. *The Ancient Engineers.* Cambridge, Mass.: MIT Press, 1970.
19. Susskind, Charles. *Understanding Technology.* Baltimore: Johns Hopkins University Press, 1973.
20. Teich, Albert H., ed. *Technology and Man's Future.* New York: St. Martin's Press, 1972.
21. Weber, Ernst, Gordon K. Teal, and A. George Schillinger. *Technology Forecast for 1980.* New York: Van Nostrand Reinhold, 1971.
22. Whinnery, John R. *The World of Engineering.* New York: McGraw-Hill, 1965.

Chapter Two

HABITAT

2.1 INTRODUCTION

In thinking about the ramifications of our living environment, one runs up against many large-scale problems whose complexity and interaction make it extremely difficult to discuss them with the kind of optimism that does not tarnish honesty. A problem as relevant and complex as the one that concerns us here, examined in our present-day setting, dictates a degree of pessimism as a reflection of maturity (rather than defeatism) and drives us to search deeper for more promising alternatives. Despondent pessimism may see none. However, healthy pessimism not only expresses concern but looks for new directions and new challenges.

The first thought that comes to mind when we are reminded of the possible shortage and high cost of energy is that we have to return to the simple way of life of our forebears—each man living on his own little farm and providing for himself and his family from his local habitat. Even if we do not exaggerate we think of a much simpler and more austere living.

But mankind has been accustomed to the civilities of living with ready access to a multitude of experiences and material goods. So on second thought we would prefer to move forward despite the problem, by a more judicious use of our resources. We can learn to conserve and share, particularly if our experiences can be richer. After all, suburban living was a compromise, brought about by the automobile and cheap energy, for us to keep one foot on the farm and the other foot in the city. Our suburbs often have names with village tagging at the end. We had grafted together the best of the two worlds. The energy problem will knock us off the fence but right on the course that was followed before the arrival of the automobile and the great sprawl of the inner city and its subsequent decay. The problem is how to do it gracefully and get the best of all the possible futures. Thus our discussion of habitat has a specific aim and direction to go. We need not act panicky and look for a way to get off by going back. There is a clear path ahead supported by historical evidence, population growth, cultural trends, and technological developments.

In the absence of a great catastrophe the direction points to more city life and more cities. But what kind of city and what kind of life? That is the question we shall examine in these pages. Let us start by sketching the origins and changes in human settlements.

2.2 ORIGINS AND DESTINATIONS OF HABITATION

To the wandering Bedouins herding their sheep and goats, the whole earth where their domesticated animals graze is a mobile habitat. Their way of life reminds us of the modern gypsies who wander from city to city to earn their living without close identification, or of modern western man with his housetrailer moving from job to job and sometimes feasting on nature's sites and monuments. Our forebears had to form settlements where they could live in large numbers to protect themselves against raiding marauders who came after their herds and women. In the jungles of the Amazon, primitive people, without knowledge of farming, lived (and still do) in villages and ate plants growing in the jungle. By contrast, other tribes, as some in Africa today, hunted for their food. As a more self-reliant way of living, man developed agriculture and animal husbandry and learned to work close together and to trade and exchange. But now the raiders came to take the land and the weaker group moved to new land. As the arable land was used up, they had to take a stand. Man specialized: some tilled the land while others trained to fight and built castles for a strong defense. So long as the castle was fortified and strong, they could all gather there and sit it out until the enemy was worn out and left. Even before that, in Cappadocia (central Asia Minor, now Turkey), people built their amazing habitat in rocky mountains, stored their food, and provided natural defense. If an attacking enemy succeeded, he demanded tax on the land. Out of this emerged dominant groups and eventually nation-states. In time it became cheaper and more efficient to trade than to fight. Cities developed around the castles and at the mouths of rivers and other natural centers of trade. The Industrial Revolution brought more people to the city; their activities increased and diversified.

Today urban living is on the rise. Living in towns and cities has developed an inner richness. Art museums, sports stadiums, music halls, and shopping malls thrive. There is a clear realization that cities and civilization are inseparable. A great city should handle its garbage and its art equally well. While at first men lived closer simply to survive, now they live together because of the excitement they could bring to each other through their myriad talents.

The city became an arena for creativity and appreciation. Man's knowledge of his talents and potentials unfolded in the city. While ancient man worked internally and kept looking out with fear, now man has come so close together that he looks inside to the city—sometimes with fear, but mostly with pleasure, satisfaction, and fulfillment.

In the city, men invented machines that they learned to exploit in agriculture. So the number of people in farming throughout the world has begun to shrink. As machines move out to the farms to do the work of hundreds of men, the men have been released from their labor and have moved to the city, crowding it and demanding a way of livelihood out of its multifaceted activities. The city grew, and affluence brought about suburban living, where people had the dual advantage of proximity to the city and its life, and of a rural living environment.

At the beginning of the twentieth century only a trickle of 15 percent of the world population lived in cities. By 1960 this figure rose to 33 percent and is projected to reach a 50 percent flood by the year 2000. By 1900, 11 metropolitan cities had one million or more inhabitants; there were 75 in 1950, and by 1976 the figure had skyrocketed to 273. Of these, 147 are in less developed countries. By 1985 there will be 17 cities of 10 million or more, three of them in developing countries, with Mexico City reaching 18 million, just behind New York and Tokyo with 25 million each. Urban areas will need to absorb 1.1 billion people in the next quarter century. There is a great irreversible momentum to the urbanization process. Today we have nearly two thousand cities whose population is over one hundred thousand, and hardly anyone believes that man, now that he knows better, will return to earn his living by toiling with his hands. Besides, of the two hundred thousand additional people in our world each day (70 million a year), a large number will live in cities (see Figures 2.1 and 2.2).

A very interesting book called *Reshaping the International Order* is the

FIGURE 2.1

PACE OF URBANIZATION IN THE WORLD,
YEAR 1 TO A.D. 2050

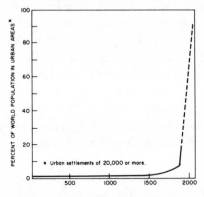

*Urban settlements of 20,000 or more.

SOURCES: *UN Report on the World Social Situation.* 1957; *The Future Growth of World Population.* UN, 1958; *Provisional Report on World Population Prospects.* UN, 1954; *Population Bulletins.* vol. 18, no. 1, and vol. 16, no. 6; Population Reference Bureau: "The Origin and Growth of Urbanization in the World"; Kingsley Davis, in Harold M. Mayer and Clyde F. Kohn, eds., *Readings in Urban Geography.* Chicago: University of Chicago Press, 1959.

FIGURE 2.2

CHANGING PROPORTIONS OF RURAL AND URBAN POPULATION
OF THE UNITED STATES DURING THE COURSE OF ITS
NATIONAL AND ECONOMIC DEVELOPMENT

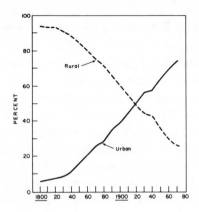

NOTE: The definition of "urban" here is that of the Bureau of Census, which in general includes inhabitants in places of 2,500 population or more.

SOURCE: *United States Census of Population.* 1950 (PO(1)1A), U.S. Bureau of Census.

outcome of the work of a team of twenty-one scholars working under the Nobel laureate in economics Jan Tinbergen, who conducted a study from 1974 to 1976 for the now famous Club of Rome, concerned with global problems and their solution. The study points out that the world's urban population has doubled in the last twenty-five years and will double again in the next generation. By the turn of the century, 3.2 billion (half the world's population) are expected to live in cities and towns, and two-thirds of them will live in the Third World. The study says, "Traditionally, the size of the city has served as an index of 'progress' and 'development,'" but that will not be true in the future as impoverished rural residents continue to flood urban areas. "If city size was ever a satisfactory index of prosperity in the past, it is barely so today, and will certainly not be so in the near future."

We agree with this observation and believe that city size should be limited. We said that we do not believe one can reverse the tide of history and send people back to the farms to grow their own food. It can be done much more efficiently and cheaply with machines. Besides, most urbanized people have no taste for working the land. They have become accustomed to more intellectual occupations. Also, trade and competition can only take place in the city. The city we must have, but we must improve the quality of its life and its functions. We do not consider the scattering of people into little towns practicable, though this will happen for a while. Soon these towns will grow and have the same problems that cities now have. But we continue to move ahead as if planning is of no immediate concern, letting things happen the "natural way." There is no easy natural way to deal with complexity.

William Marlin, in *Saturday Review* (August 21, 1976) has given us a very lively view of the transformation of our cities which I consider a temporary "solution." The American city is in a process of comeback. People were "moving by the millions from urban centers and outlying farmlands, passing like ships in the night of suburbia." They are now back, "impelling renewals of the same environments from which those who had money fled and towards which those who had little money gravitated." Bright cities have sprouted in the farmlands. There is "resurgence of neighborhood pride" in cities like New York, Boston, and Philadelphia, "a rediscovery of the quality and character of long-abused streets and parks." Theatres and many new and refurbished restaurants are thriving. Discovering that resources are finite, inner cities have made the most of what they have by recycling old bricks and front stoops, and by infilling with new construction. Thus, for continued life, old buildings have been recovered and row houses spruced up. Some old cities, like Boston, are "a lot more swinging and outspoken than [they] used to be." The Galleria complex of Houston "mixes stores, shops and hotel facilities around a cavernous skylit interior mall." The new Crown Center in Kansas City is becoming an unpretentious magnet for cultural life. Other cities are landscaping riverfronts, mixing cultural, educational, and business activities, and even restoring leafy residential districts with redevelopment programs for social amenity and visual beauty. Most are savoring interaction of people·and purposes at street level.

Habitat has grown from the Bedouin's tent and the tribal village to the metropolis and the megalopolis. Dinosaurs are said to have perished because of their immense size and declining resources (food). More adaptable and less gigantic animals have survived and preserve many of the good qualities of their extinct forebears. There are signs of decay in our large cities. There are also eleventh-hour attempts to revive them.

The city now has a large number of people who must adapt to its life and who may not have the skills necessary for survival. When there is a juxtaposition of the very poor and the rich, the former, now observing first-hand the gap between having and not having, feel despair and may turn to crime. Those who have the money leave, depriving the city of their resources to rejuvenate itself.

Most cities today are losing housing and population, especially middle and upper income brackets, leaving the city with a population that de-mands more in services than revenue allows. New York City as of 1977 had lost six hundred thousand jobs in twelve years. As a result, there is a higher ratio of renters to owners. The real estate investors' cost—land, materials, labor, operating expenses, taxes—adds up to more than a typical consumer can pay. A paradoxical situation arises, with very few vacancies and investors going broke. Between 1970 and 1975, 170,000 units were taken out of the rental market. Actually, this may be a good way to deal with the problem of declining population and housing decay. As parts of a system become old and obsolete and the system can con-tinue to function without them, they can be removed without harm. An eyesore is less damaging socially than a nagging economic sinkhole of funds for which there is no hope of recovery.

One way to assist the lower-income groups is through subsidy. Incen-tives from the government to the private sector are needed to supply housing for would-be lower-income renters. In St. Louis, the tenant was able to afford about a third of the cost of operating his unit. Under current guidelines one and a half million are eligible for housing subsidies.

Another patchwork solution is to invest in housing rehabilitation, which costs about $50,000 per unit with union labor and about $30,000 without it. This is too high a price to pay, and there is no need for it, due to the declining population in the cities. A national policy toward rehabilitation would, however, benefit the construction industry. The third option that has been proposed is to train people and involve them in taking charge of their environment. Tenants have begun to operate their own buildings when landlords have not provided adequate service. The worker on rehabilitating structures in the process of tax foreclosure is given the option of contributing his labor towards buying the unit. The program is seen as a way of contributing to the solution of housing and unemployment problems.

To maintain its viability, the city must be supported by a productive population. The costs have risen with the pressure to clean up the environment and with the shortage of cheap forms of energy. The human and the environmental or habitat problems have run up against each other. Facing up to the problem is being delayed a while longer.

The Urbanization Route

Tents and mud huts grew to hamlets and villages and the villages grew into towns. The town grew to a city and the city became a metropolis. The metropolis became a megalopolis and the megalopolis became an ecumenopolis. Some metropolises simply sprawled outwards like Los Angeles and others like New York spread upwards.

The outward-growing city gobbles up precious land and needs vast streets and transport systems. The upward city is gradually taking the shape of a megastructure although its piecemeal growth has impeded its functions and its enclaves have compartmented its life. What we need is better planning and design of our megastructures. That is what Compact City is about.

2.3 AN ANTERIOR ASSIGNMENT

You are a well-known architect and city planner, living at the turn of the nineteenth century. You have been asked by the mayor and city council of Manhattan to help by studying a film of the city as it will be in the last decades of the twentieth century. What you see are skyscrapers casting long shadows over roads and buildings; avenues and streets crisscrossing in a never-ending web, elevated roads between buildings, subway entrances and subways, railroad stations, bus terminals and airports, roads leading in and out of the city through tunnels and bridges, and a myriad of other things which do not seem to have rhyme or reason. Life seems to be taken as movement from building to building. An individual who lives in one tall building walks to an elevator and takes it down to the street level or to the level where his car is parked. He drives out to the street where he joins a mad noisy rush of traffic whose hum and din can be heard for miles, stopping and speeding, screeching and colliding, smashing and blowing horns, parking in the middle of the street or climbing the sidewalk to run over pedestrians and smash into buildings out of control. To be "safe" the individual may walk or take a taxi to a subway station where he walks down several stairways or takes escalators to the subway platform where his ears are bombarded with the sound of iron on iron of passing trains which stir up the dust and blur the vision. When he finally takes his train, it speeds him on his way, jerking and shaking him as it stops and moves from station to station. At the rush hour he is squashed by other passengers. Arriving at his destination he may go through a mirror image of movements to get to his office, to a friend's house or to a shop. Later he has to return to his apartment in the tall building going through the same agony all over again. You also observe that everybody is doing the same thing at the same time of day, both to work and to play and to rest, and you note the pollution, congestion, excessive hustle and bustle which make it difficult for anyone to be aware of anything except to manage to "get there" wherever that may be. The sky, the sun, the rain, the snow, and the wind are incidental and can be more of an annoyance than help. To get out of the city to nature is a thing so difficult, expensive, and time-consuming that few people think of doing it. Late into the night, city life quiets down and now the streets and office buildings are empty and the apartments and houses are full and still. Turning to the living conditions, you note the energy consumed in heating and cooling the city, its water shortages, its scattered sewage plants polluting rivers and ocean, the nearby garbage dumps and junk yards.

You are asked if New York could be designed for better living in the last decades of the twentieth century, but you have to accept the population problem and must use the third dimension. The awkward inefficiencies should be minimized, life must be more livable, nature more available, polluting facilities eliminated or distant, transportation minimal, energy consumption no more than it is. What would your design be like?

For those who worry about energy use to move elevators or to pump water upward, we note that one-fourth of all the energy used in our society is consumed by transportation, and the bulk of transporting occurs in the city where millions (80–90 million) of vehicles move short distances many times a day. The total mileage of such short distances of commuting is astronomical and uses an inordinately large amount of energy. The design of the new city should minimize the distance and time of travel.

Los Angeles with its fantastic sprawl is at the other extreme. Little can be done in Los Angeles without the automobile and substantial travel time. That much travel is the major cause of pollution. "Do not take deep breaths" the signs read in some parts. How can a city that has so much sprawl be put together in a design that offers people private residences with gardens, at the same time saving their health, time lost in transportation, and money spent on cars, and offering accessible cultural and social life for those who want it? These are things that can be remedied with great difficulty or not at all in the existing approach to our problems.

The sprawl of cities with their road systems often covers hundreds of miles. Most cities are built over land that is fertile and could be cultivated. Certainly, the outlying industries and their polluting activities often extend to the surrounding land and water. Land occupied by a city like Washington, D.C., covering an area of 100 square miles, if turned agricultural (at 640 acres per square mile and three people fed by the produce of one acre), could feed $100 \times 640 \times 3 = 192,000$.

Now use of the third dimension would release most of this agriculturally usable land. We have not yet reached the point where we are excessively concerned about the availability of agricultural land, but soon we will, as indicated in Table 2.1. Making good use of land is rapidly becoming a concern.

Another problem is the use of energy in transportation and heating in the winter, both of which account for nearly 40 percent of the total energy used.

Most of the social and cultural amenities of the large city are not easily accessible to its population. Nor can a small population, e.g., less than one-half to one million, support such activities. It would be desirable to make it possible for all the people to have access to the entire city. A

TABLE 2.1

AGRICULTURAL LAND RESERVES
(in millions of acres)

Region	Ultimate maximum arable land	Land in the cultivation cycle	Land harvested per year
North America	968	543	274
Western Europe	383	314	220
Japan	20	15	15
Australia, etc.	371	143	47
Eastern Europe including USSR	944	691	477
Latin America	1060	316	190
North Africa and Middle East	212	131	72
Main Africa	1045	413	180
South Asia	687	662	581
China	301	292	247
World	5991	3520	2303

city's life-support systems—clinics and hospitals, churches, schools, shopping areas, water, electricity, sewage and garbage services—need to be convenient and reasonably priced. They should be well integrated with the other systems. Sports, gardening, and other interaction with nature must be available and near. City life should be safe and loitering for mischief should be made more difficult.

2.4 THE INTEGRATED CITY (COMPACT CITY)

There is a school of planners and demographers that argues that the trends today, rather than favoring the high-rise metropolitan city, in fact tend towards decentralization to small towns scattered throughout the country. It is said that the reasons for the existence of the central city are no longer there. The need for the city as a transportation hub has been removed by the automobile and the highway; labor is highly mobile and is no longer dependent on heavy concentration of the population. Banking, whose money draws business, is now widespread to the smallest communities. Television and not-too-distant three-dimensional picture communication will make actual theatre and museum attendance obsolete. Small cities can be just as sufficient and satisfying as large cities, minus their frustrations. Our rapid intercity transportation systems can move materials in and out of the city. Communication has made it increasingly unncessary to travel or move about to see other people. The presence of local supermarkets and shopping centers, satellite clinics, and nationwide divisions of business and industry with their jobs (perhaps shared by neighboring towns) diminish the need for living in the large city. Among the advantages of the small town are fresh air, sun, little or no crowding, easy access to nearby parks and recreation areas, and a more desirable social life with everybody acquainted with the people of the town.

Some ecologists prefer the linear city, stretching along a line to alleviate the intensity of pollution on the inhabitants. At least in principle each house, rather than being surrounded, is only bordered on either side by other houses. It is also surrounded with parks and water bodies for recreation and to absorb falling pollution particles.

The challenge to this author was to transform the science-fiction ideal of Compact City (proposed by some in the spirit of H. G. Wells) to a concrete, livable, and very possible reality. It was by working with urban designers interacting with civil engineers and social scientists at places such as the Urban Institute in Washington and studying the voluminous literature on new architectural designs that the parameters and their feasible values were developed. A few examples are the physical characteristics of the city, formalization of the space and time principles, gradual implementation through modular construction, aesthetic considerations including plants and fountains and malls, the effect of Circadian rhythms, and safety features that needed to be built into the city. One of the major problems has been how to arrange for maximum sunlight in the city. Some early versions were proposed as windowless closed structures, which drew strongly negative reactions. Thus the city emerged as a living entity with its excitements and problems out of a welter of modernistic and far-out hunches.

The idea that three-dimensional cities can be better built is not new. Not only the people who live in such cities have their idea on how they could be improved, but well-known and dedicated architects have considered the design and some, even, a small-scale implementation.

Paolo Soleri, who was born in Italy, has lived near Phoenix, Arizona, occupied with building his city called *Arcology,* which is a three-dimensional megastructure, a miniaturization to compact form in an effort to reduce time and space obstacles. His idea is that cities acquire greater efficiency by using the third dimension. To Soleri, complexity and present city size make it the only solution.

Constantinos A. Doxiades promoted the idea that the whole wide world would one day be a large interconnected urbanization. In Doxiades' consideration of cities this was the ultimate stage for the city. In it he found both the real and the ideal form for a city; he called it *ecumenopolis.*

In a more practical vein, Montreal commissioned Moshe Safdie to design *Habitat* (a megastructure located in the outskirts of the city) involving modular units that are stacked together to form a community. To him the livable cities create choices not possible in lower-density environments.

There are other conceptualizations of the integrated three-dimensional city which we do not mention here and which the reader will find mentioned in *Compact City* (3).

Two principles lead one to the consideration of compact cities. They are the *space principle* and the *time principle.* Briefly, the space principle says that with the rising population, increased activities, and the need for efficiency in the use of land and energy, material, and time to get there, we must make use of the third dimension. The time principle says that we do not all need to be up and doing things at the same time. We could align or stagger work and play hours to make it easier for all of us. Our office buildings, highways, and public facilities which consume our resources lie idle most of the evenings, and their size has to be large to accommodate maximum demand. For example, one table in a restaurant could seat people round the clock while many more tables would be needed if everyone went to eat at the same time.

We could economize in the use of the world around us and still satisfy all our needs. It is a matter of getting used to the idea. The rural population who move to the city take time to adjust to its crowding, noise, pollution, and the like. But they like to be there because its opportunities are greater and its social rewards, despite many weaknesses, are also greater.

Compact City is a three-dimensional city occupying an area of land equal to about nine square miles and rising upwards sixteen levels separated by thirty feet. Each level is cut through its center by thirty-two radial streets. The level also has ring-shaped streets growing outwards. The main activities of the city are at the core near the center. Kindergartens, small shops, and churches are interspersed among residences in the outer areas. The city's total living area is nearly 140 square miles. Its malls, plazas, and stadiums cut through two or three levels. Two million people live in the city (whose density is fifteen thousand per square mile, less than most large cities) and when their number increases some move on to live in other cities. The city may be started in the outlying suburbs of existing cities to make a slow transition to it. First its central working-and-other-activities core is built and gradually expanded to include living areas. Compact City is served by a system of elevators that are conveniently located. Its roof is a garden with large conservatory-style greenhouses. The city is surrounded by parkland with a highway system partly underground to other cities and with a transport system to an industrial park where heavy industries are located and to a nearby airport. Life within Compact City has most of the amenities of large cities but without many of their problems. Walking, electric cars, and a rapid transit system on its eighth level provide all its transport needs.

Access to nature is rapid through approximately one thousand exits from its inner roads to the outside and to its roof by elevator.

People will live in individual two-floor housing units built from durable, nonflammable, soundproof colorful materials, as there is no threat from nature. The partly modularized building materials may be moved around to alter the design. The city will be lighted partly with sunlight but mostly with artificial lighting. It will be cooled, as its activities and people will generate considerable heat which raises the temperature. In the winter, cold air would be blown from the outside, and in the summer, depending on its location, it may need air conditioning.

Our study of Compact City indicates that if Manhattan or any similar three-dimensional city had taken the Compact City structure, nearly all its hazards would be far less than they are now.

Walking in Compact City would again become fashionable, as the maximum distance would require about forty-five minutes of normal walking. Transport in each level is by electric car and there are no internal combustion engines. Ten thousand such cars are needed for the entire city, whose activities could be extended round the clock. The rapid transit system on the eighth floor requires two hundred cars and involves minimal waiting. The cost of transportation is negligible and people do not need to own cars. They can rent them for intercity travel. Thus there are no traffic jams in Compact City. Also, there is no pollution in the well-circulated air; noise pollution is minimal, particularly away from the business districts. Material and food deliveries, water, electricity, ventilation, and sewage and garbage removal systems are all carefully designed with multiple reliabilities. Although there are no perfect fail-safe systems, those of Compact City would be better planned and designed as overall systems than their piecemeal growth in conventional cities. Neighborhood and work-center concepts have also been considered in the design. The most exciting aspect of Compact City is the accessibility of all its urban opportunities to all the population, whether for work, pleasure, social gathering, political participation, or planning city life.

There are several problems to consider, of which five appear to be serious at first:

1. The cost of the city and how to bring it about.
2. The availability of energy to sustain its activities.
3. The threat of tampering with the life systems of the city.
4. The threat of regimentation arising out of city design or social and political control.
5. The social life of the city—whether it would affect people's sanity and happiness. Can the structure of the city serve as ground of fighting between people occupying different levels and different neighborhoods? How about opportunities for the disadvantaged?

We do not have sufficient room to cover all these questions in detail, but we shall touch on most of them.

Cost

The annual costs of our transport systems today boggle the imagination. They are nearly $1000 per vehicle per year. Road construction and repair use considerable amounts of materials (recall that 40 percent of Los Angeles is taken up by roads and in most cities it is near 25 percent). Then there are the costs of buying new vehicles and maintaining them, building garages, and loss of life and productivity due to accidents. In fact, the cost of all these and related activities not needed in Compact City is staggering and measured in billions. We have estimated that savings from transport system maintenance and expansion can go a long way to make the cost of building a compact city attractive. Also, people have to live and work somewhere and the money spent on building construction can be diverted to Compact City. Our calculations show that the costs of living in the city will be lower than at present for higher-income groups and higher (as one might expect) for lower-income-earning families.

One of the most attractive features of this type of urbanization is the large long-range opportunities it will create for employment in new industries oriented towards this new style of living and the challenge offered by planning the interior of the city with water fountains, lighting effects, public gathering places, special parks, and interior plant industries with potted plants that can be removed, recultivated, and delivered.

The interior design of the city has been discussed in great detail in the book *Compact City* (3) and will not be repeated here. We strongly feel that it will not be long before the full challenge of a well-designed integrated three-dimensional city will be facing us. The historical trend has been in that direction, and despite their shortcomings, cities are popular and exciting.

Energy

Three separate exploratory studies were made by the author and his energy students to calculate the energy needs of Compact City. They took into consideration the time of year, interior lighting, air conditioning, electric appliances, transportation, operating light industry and manufacturing based on electric energy, and even the removal of the heat generated by people and by appliances.* Without giving the full details here one of these studies showed that Compact City can save from 23 to 35 percent of the energy consumed by a comparable-size conventional city. For example, even though its lighting requirements would be considerably higher, the energy consumed does not match the nearly 25 percent of energy we consume in transport, most of which is expended in city traffic. Energy consumed in winter household heating would not be needed, but Compact City would require air conditioning, although its streets and homes would not be as hot in the summer since they are not exposed to the sun. As far as efficient energy utilization and saving, Compact City offers a very attractive alternative to the conventional city.

In recent years we have turned to the use of solar energy for household heating needs. The effort in this direction is accelerating to the extent that the U.S. government has considered implementation of solar heating in 2.5 million houses by 1985. As its technology becomes less expensive, solar energy may be used in a variety of ways for compact-city purposes. The sun provides 1400 watts per square meter (4,730 Btu per square meter per hour) in outer space but an average of 160 watts per square meter on a horizontal surface in southern New England, and if the surface were to rotate perpendicular to the sun's rays this

---

*Refer to chapter 3 for further information on energy.

figure would be about 300 watts per square meter. The total electric power demands of the United States could be supplied by solar energy plants with a total area of 2000 square kilometers (0.03 percent of total area and 2 percent of land area given to roads).

High-temperature collectors can be used for generating electricity by concentrating the solar energy from a large area into a small focused collector. Thus, for example, the energy system for the industrial park associated with a Compact City may be designed for optimum utilization of solar energy.

Threats

Although we do not wish to dismiss them lightly, our concern about threats to Compact City are the same as those we have for existing cities. Wherever we see easier threats to Compact City, we also see a possibility for counteraction. Midtown Manhattan could be thought of the same way. What threats have there been to it and how will it survive imagined threats in the future?

The Economic Situation

The thought of how complex, difficult, and intertwined our problems have become has turned people into intellectual pessimists, but many remain viscerally optimistic. Nowadays we are all painfully aware that resources may be increasingly short and expensive in coming years. As somebody pointed out, children are competing with travel, the new house, and professional standing. Once the checkbook is balanced and all other desires have been indulged, a couple will think of having a child, or at least having a child will rise on their priority list. Zero Population Growth (ZPG) has worried some people about the social and cultural implications of a markedly older population. By the year 2000 there will be twice as many people over the age of 65 (43 million) as there are today. Their social security, pension, and medicare needs will exert great pressure on the economy. Of course raising the working age to 70 alleviates the financial burden of old age for the government. Increase of older people could mean less innovation, and fewer people available to attend to, care for, feed, and house a larger number. Many futurists believe that ZPG will reduce pressures on the environment and resources and would probably increase per capita income. It will give society the opportunity to invest more in the quality than the quantity of life. Air, water, and noise pollution would be reduced when families with fewer children have more discretionary income to spend on the pursuit of pleasure—and for better health and education. With an older and less adventurous population there would be less pressure on all the natural recreational resources. Two- and three-bedroom units are in demand for the smaller family rather than single-family houses. Education should improve as the teacher-student ratio improves. There will be continuing and increasing demand for adult education. Children will be healthier as there are fewer of them in the family. The joy of having children will be enhanced. A controlled population will allow the re-ordering of priorities and reassessment of values before they are dictated by scarcity.

Social Life

Although living in Compact City can be thought of as it is in any modern metropolis with its delights and major problems, one is inclined to make greater social demands on Compact City than there are now in any of our cities. Life in a conventional city has developed piece by piece as people joined to live. In most neighborhoods today, many people don't even know who their neighbors are and they don't feel a need to. But for Compact City our reasoning may go something like this: if we are going to go this far in designing a city to meet the total physical needs of man, can we also design its functions so that his social needs of responsibility, cooperation, enthusiasm, and adherence to the law are enhanced and become natural daily expressions?

Short of getting people to cooperate willingly to improve social living, there is no social miracle that will take place by building Compact City. It is easy enough to mold natural resources to build the city. It is not reasonable to expect that by simply manipulating our physical resources we can learn to live together as better human beings, despite the fact that it may give us more personal time and reduce our frustrations. It is not yet clear to us in the twentieth century how to make it possible for people to mold themselves voluntarily to integrated social living. Not much of this type of life has been attempted in our time on a massive scale. The Chinese experiment, which is considered a success, initially gave people strong direction to get them to cooperate in forming a new society. Another example is the Israeli kibbutz, which by and large has also been a successful experiment. In some European countries people are traditionally engaged in governing their affairs.

The social life of Compact City would be more satisfying if people from age fourteen on were to work together to maintain the life functions and administration of the city. Some will be more expert at it than others, but apprenticeship will be a part of its opportunities. For this participation some will get their services free, and others will have their taxes reduced.

One would not have to coerce participation if one succeeded in making it a policy for Compact City, advertised it, and pursued it from the top down to engage people. The system of living in Compact City should be well understood and broken down into its parts to make it possible for volunteers to put in small amounts of time.

In a democracy people pay taxes to support their government. Tax money comes from people's labor earnings. For most, there would be little difference if they labored to pay taxes or labored directly as a public service in the governance of the city. In addition, the social good requires that people pay taxes. The idea of forced indirect participation in government through taxes may seem coercive in a free society, but it is here to stay. In fact it seems more democratic and responsible for people to volunteer their time.

How to Make Compact City Socially Livable by Designing

Consider each level as being divided into six sectors, and light the sectors by operating stores, etc., in each sector on a four-hour stagger. Coordinate this schedule between levels, so that all sixteen sectors on the same vertical axis are on the same schedule.

Advantages: Families are much less likely to suffer from scheduling problems. Use of central facilities is still averaged over six periods on each level. Neighborhoods are sufficiently dense with same-schedule people to justify neighborhood schools and social activities. The small-town atmosphere is present, since people on a level divide naturally into sector-towns of roughly twenty thousand. Interlevel rivalries are lessened, since each individual belongs to *two* classes with distinguishing characteristics: one horizontal, and one vertical. For example, social activities between "towns" on different levels can be easily encouraged.

Disadvantage: Use of local facilities is not averaged. But at least, people only have to work at local stores at times when local people are awake. A night-cycle will save on lighting and energy expenses in the evening (and may be of psychological value as well). Essential services can be concentrated on those sectors most likely to need them at a given time of day.

General Discussion of the Social Order

To try to understand what motivates human beings to opt for new residential arrangements we can look to suburbia. Here is a laboratory for investigation. In the late 1940s, Congress passed legislation that made the acquisition of home loans for veterans a reality for single-family dwellings. More encompassing FHA legislation followed.

Why did Americans turn to this new life style with such a vengeance? A large set of circumstances coexisted—feasibility, the postwar baby boom, the wide-open-spaces syndrome, massive highway construction, and the cultural makeup to accept and desire change. Can we find a similar desire for a new form today?

Making Compact City a reality may be economically efficient, and at the same time it identifies a need or desire for a new environment on the part of a large segment of society. A set of conditions must be determined that will make dwelling in Compact City an aspiration in the context of the American experience.

Cities as we have come to know them are composed of a highly integrated set of interrelationships. In the city there exist, simultaneously, specialization and complementarity. A juxtaposed set of mixed cultures creates an atmosphere of dynamism that is conducive to creativity. This circumstance is often a necessary condition for fostering the birth of innovation, especially social and cultural innovation, which is a result of human creativity. This creativity is what allows us to recognize the need for change and to make more appropriate response to problems, be they social, cultural, environmental, or sociotechnical.

A potential problem is the possibility of a clash between a newly created superenvironment and the cultural and intellectual diversity of American cities as they have evolved to date. The danger of homogeneity in the resultant population of Compact City may stunt creativity. (The urban minorities in the American experience have been overrepresented in terms of contributions to the arts, music—both serious and popular, literature, and the social sciences.)

A second problem is the possibility of reducing choice by comprehensive planning. This may be a short-term blessing for the hassled urban dweller who is confronted with more diversity of action or opportunity than is desired, but that kind of utopia is short-lived. Some people will accept only presently existing alternatives. This is usually the result of a less-than-useful educational experience and/or complacency. However, a significant minority of the population will demand a construct that will allow for shifting methods and priorities in light of existing situations.

On Creating Equality

There is a long-run cost associated with maintaining inequality. The value of equality has some inherent appeal to planners. However, it is a problem on which more than passing comment can be made. It involves the reordering of an established set of social *values*. That the optimal mix of values has not been achieved is hardly a point for lengthy discourse. But a good deal more thinking must be done by social scientists on the mid- to long-range ramifications of imposing values. There are many cases today in society where the imposition of values has occurred. What are the consequences? Did you ever meet a military officer from over- seas who had received much indoctrination and training by the U.S. military? How about a six-year-old who has been exposed to a steady diet of television since the age of three?

In a society, status differentials will always exist. If everyone had an equal share of material wealth, what would be the new indicators? Would we, as a society, turn to channeling this status-achieving effort into more humanely productive terms?

One central point is clear. There is no costless way to eliminate in- equality in levels of income. The process of doing away with poverty, an essential goal, will have both economic and emotional costs. The Ameri- can experience presents no example of this kind of concern or benevo- lence. To raise the expectations of suppressed groups (such as the inner- city poor) to soaring levels with the promise of overnight remedy by relocation in new physical structures is a game that was played on a selective basis during recent years. The results were seen in new aban- doned and/or destroyed housing in St. Louis, New York, and many other cities. A whole complex of social arrangements must be made to allow groups to have the opportunity to make self-gain. No simple physical remedy exists because the values and attitudes will be missing.

If Compact City were alive and well today, what would be the conse- quences? Would a significant re-alteration of the meaning of community occur for the good, or would it merely be a differently shaped Levit- town? The answer depends on putting higher priorities on social goals and social planning than on material achievement.

There is a story of a man who used to say to his children who complained about trivial shortcomings in the cooking, "When people are really hungry they will eat stones."

A nineteenth-century architect would have gravitated naturally to Compact City design for Manhattan, but it would have been resisted because people did not have to accept it then. It was not their problem, but it is ours multiplied many times over. When the need presses hard, thinking and cooperation go hand in hand.

So long as the pressure is not so great as to make us really uncomfortable, we can continue to act as children in a fantastic toyshop. The future will not take care of itself as it used to (if indeed it ever did). There are too many of us and our demands on nature are reaching the breaking point. What are the promising alternatives ahead and for what style of living? That is the question the reader must try to answer for himself.

2.5 EXERCISES

1. Jules Verne wrote to his father, "Everything one man is capable of imagining, other men will be capable of realizing." Indicate the extent to which this statement is true and give counterexamples.

2. Describe and contrast the kind of cooperation involved in implementing the space program, the kibbutzim, and the Chinese social experiment. Take into account the backgrounds of the societies.

3. Western civilization has been wedded to the analytical method of science. Consider the Eastern method of contemplating the "whole" as exemplified in Zen Buddhism. Suggest ways of incorporating that method with the analytical method to improve the prospects of coping with problems.

4. The world's population is projected to double from its present 4.5 billion level in about twenty-five years. Give three innovative suggestions as to how the Third World countries can contain this rise in population. Describe the Chinese experiment and how it has coped with the problem.

5. By comparing the volume of Compact City with some standard buildings compute energy requirements for its nonindustrial activities.

6. Discuss technological spinoffs from our space exploration program that would be of use in Compact City: recycling, miniaturization, material design.

7. Design a horticulture system to keep Compact City interiors supplied with green plants. Most plants need outside sun part of the time.

8. Give a program for people to cultivate vegetable gardens in the surrounding areas of the city. Recall that these areas are nearby and quickly accessible. They can also use heat rejected from the city in the cold season.

9. List the opportunities of a successful Compact City that point to greater social values and growth than there are in a conventional city.

10. Prove that the opportunity exists today for terrorism to destroy skyscrapers in our large cities. Give three reasons why no such action has taken place. What is its likelihood in the future?

11. Estimate the size of concrete manufacturing industries, their location near the construction site, and labor opportunities to build Compact Cities. What effects would such massive construction have on the economy, and how can they be met?

12. How would you solve the dog-walking problem in Compact City? Maybe your solution could be implemented in New York.

13. Find out the potential of using fiber optics to bring sunlight to the interior of Compact City.

2.6 BIBLIOGRAPHY

1. Brown, Harrison. *The Challenge of Man's Future.* New York: Viking, 1956.
2. Chase, Stuart. *The Most Probable World.* Baltimore: Penguin, 1968.
3. Dantzig, G. B., and T. L. Saaty. *Compact City.* San Francisco: W. H. Freeman, 1973.
4. Drucker, P. F. *The Age of Discontinuity.* New York: Harper & Row, 1968.
5. Editors of *Fortune. The Environment: A National Mission for the Seventies.* New York: Harper & Row, 1969.
6. Elias, C. E., Jr., J. Gillies, and S. Riemer. *Metropolis: Values in Conflict.* Belmont, Calif.: Wadsworth, 1964.
7. Ellul, Jacques. *The Technological Society.* New York: Knopf, 1964.
8. *The Futurist* (magazine). World Future Society, P.O. Box 30369, Bethesda Branch, Washington, DC 20014.
9. Gunther, John. *Twelve Cities.* New York: Harper & Row, 1967.
10. Jacobs, J. *The Death and Life of Great American Cities.* New York: Vintage/Random House, 1961.
11. Jacobs, J. *The Economy of Cities.* New York: Random House, 1969.
12. Meadows, Dennis, et al. *The Limits to Growth.* New York: Universe, 1972.
13. Mesarovic, M., and E. Pestel. *Mankind at the Turning Point.* New York: Dutton/Readers Digest, 1974.
14. Mumford, Lewis. *The City in History.* New York: Harcourt, Brace & World/Harbinger, 1961.
15. Skinner, B. F. *Walden Two.* New York: Macmillan, 1948.
16. Tinbergen, Jan. *Reshaping the International Order.* New York: Dutton, 1976.
17. Toffler, Alvin. *Future Shock.* New York: Bantam, 1970.
18. Tunnard, Christopher. *The Modern American City.* Princeton, N.J.: D. Van Nostrand, 1968.
19. Ward, Barbara. *The House of Man.* New York: Norton, 1976.
20. Wilbern, York. *The Withering Away of the City.* Bloomington: Indiana University Press, 1964.
21. World Future Society. *The Future: A Guide to Information Sources.* (See ref. 8 for address.)
22. World Future Society. *The Study of the Future: An Introduction to the Art and Science of Understanding and Shaping Tomorrow's World.*

Chapter Three

ENERGY AND SOCIETY

3.1 LIFE AND ENERGY

In his science-fiction book *Last Judgement*, published some time ago, the famous geneticist J. B. S. Haldane has an announcer from Venus say: "It was characteristic of dwellers on Earth that they never looked ahead more than a million years and the amount of energy was ridiculously squandered." Science-fiction writers are often more prophetic than scientists and engineers. They frequently are among the forerunners in predicting beautiful new things as well as in foreseeing catastrophe.

As of this writing our energy is not depleted to the point where civilization is nearing its end. However, we are at the beginning of an energy crisis, because certain fuels on which we place great reliance, such as petroleum, are nonrenewable and are being depleted rapidly. Not even the vast petroleum resources of the OPEC* nations can last forever. The immediate problem that we face, then, is not that we will run out of energy, but that certain forms of energy, such as petroleum and natural gas, will become scarce and expensive. For a nation that has become accustomed to cheap energy this is a disconcerting realization and one we must face immediately: (1) We should conserve energy, particularly those energy resources that are nonrenewable. We should have more economical cars, our houses should be better insulated, we should get used to rooms that are not overheated in winter or overcooled in sum-

---

* Organization of Petroleum Exporting Countries.

mer, we should convince supermarkets to place covers over their refrigerated display cases, for example. (2) We should develop alternate energy technologies that derive their energy from the renewable sources such as solar energy. (3) We should also increase the efficiencies of industrial equipment and household appliances so that less energy is used. Large cars should be heavily taxed. (4) We should develop prudent government policies that encourage energy conservation and new energy technologies—we are in fact one of the few major nations without an established government policy. (5) We should educate our young to be energy-conscious by introducing courses on energy into the curricula.

3.2 THE ENERGY CRISIS

The media give the impression that there is a single solution to the energy problem. There is no *one* solution because the problem involves different energy technologies, a panoply of institutional arrangements, new life styles, new concepts in international politics and trade, and new standards and codes. These must all be designed, approved, and integrated.

It is equally necessary to understand that there is no single energy resource that is best. For example, transportation requires a resource that delivers proper mileage and has a reasonable range. Petroleum is one such resource, but it is not the only one: alcohol-petroleum mixtures, hydrogen, or electricity can also be used. With some additional work, these technologies are feasible but this does not solve the problem. For instance there is a vast network of gasoline service stations. The switchover would be a major undertaking and would cost an enormous amount of money.

The available resource options are complex, but they must be developed and used if we are to maintain our current energy consumption. As of this writing we are in an energy crisis. There are many who disagree but this is purely a matter of semantics. A *crisis* is defined as a *turning point*. We surely are there and we must take dramatic measures if we want future generations to survive. It is essential that we change our entire energy picture.

The crisis we are facing has many causes: political and economic factors; neglected technological development; and a life style that thrives on high energy consumption that depletes nonrenewable resources, demanding present comfort at the expense of the future.

Like water and air, energy is one of the basic necessities of life. This is true regardless of whether we talk of natural energy such as sunlight or of technological energy such as electricity. Without energy we could not build shelters that protect us against the extremes of weather. We could not heat ourselves in winter nor cool ourselves in summer. Without energy we could not grow the food that nourishes us. Industry would not exist without energy nor would transportation. Our institutions, such as government, business, utilities, or the postal service, could not function without energy. It can be said, without exaggeration, that energy is one of the basic ingredients of civilized life, not only in industrialized nations, but in the developing countries as well. The form or the application of energy may differ from place to place. The amount of energy consumption per capita may be different in different locales. Nevertheless, as a basic need, energy is ubiquitous and without it we could not exist.

3.3 ENERGY CONSUMPTION

Figure 3.1 depicts the U.S. energy consumption per capita in million Btu's (British thermal units) during the period 1850–1973. It may be

FIGURE 3.1

U.S. ENERGY CONSUMPTION PER CAPITA, 1850–1973

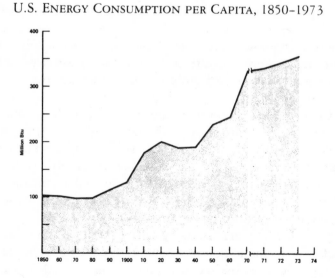

SOURCE: U.S. Bureau of the Census; U.S. Bureau of Mines.

seen that since 1850, the per capita energy consumption has more than tripled. The curve is interesting in another way: it does not increase monotonically, but fluctuates. This is due to the many factors that influence energy consumption, such as population, capital investment in industry, weather, changes in industrial complexion, war and peace. If we can change these factors we can change the rate of energy consumption. Of course, such factors must be changed carefully because they affect the economy and the society.

Figure 3.2 shows the relationship between the gross national product (per capita GNP) and the energy per capita for different countries. This graph shows that there is a correlation between an economically affluent society and the availability of energy. However, it is not clear that a given affluence always requires the same energy consumption. This point is evident from the location of several different countries in the shaded area in Figure 3.2. Thus several countries have the same per

FIGURE 3.2

COMPARISON OF WORLD GNP PER PERSON
WITH ENERGY CONSUMPTION PER PERSON

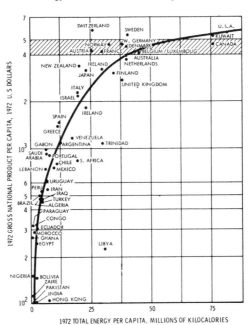

capita GNP and a low per capita energy consumption, indicating that high GNP can be attained without necessarily high energy consumption.

Although there appears to be high correlation of the data in Figure 3.2, this may be fortuitous and is being questioned in recent literature. According to O'Toole, referring to 1974 data,* a reduction in oil imports by one million barrels per day would lower the gross national income by $15 billion and would cost 400,000 jobs (5). This seems plausible as long as we do not change our energy mix but seems questionable when we agree once and for all that our energy mix must be changed.

Data Resources, Inc. (5) has constructed a different model wherein the reduction in energy is not achieved through rationing but through price increases. According to this model, cutting energy growth by about one-half would reduce the GNP only minimally. Clearly there is no single answer and econometricians themselves are not in agreement. The relationship between per capita energy consumption and per capita GNP will have to be studied in greater detail. Preliminary comparisons of the United States with Sweden and West Germany suggest that much depends on life style and the type of equipment such as automobiles and hot-water heaters. Other problems that need to be considered are the rate of productivity, and labor- vs. capital-intensive industry. All we can say at this point is that there are indeed relationships among per capita GNP, per capita energy consumption, productivity, and the type of industry (i.e., labor-intensive vs. capital-intensive) but we do not know how these relationships work quantitatively.

Figure 3.3 shows the relationship between population and energy consumption in the United States. Although the increase in energy consumption has been faster than the population growth, it is still evident that energy consumption is related to population.

Figures 3.1, 3.2, and 3.3 indicate that energy consumption is related to population and to economic affluence. There is no quantitative equation that relates energy to population and the GNP but there is a qualitative relationship. The quantitative difference are dependent on life style (for example, in Germany bedrooms are not heated and economical

---

*Accurate energy econometric data are exceedingly difficult to obtain because there are numerous sources which are inconsistent. To exacerbate the problem, data are slow in coming and are generally a few years old. These problems make it difficult to develop new policies and take quick action.

FIGURE 3.3

COMPARISON OF U.S. POPULATION
WITH U.S. ENERGY CONSUMPTION

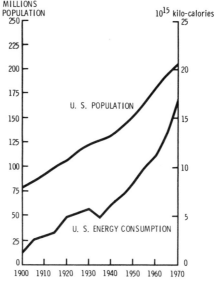

water heaters have been developed), on distances traveled, on climate, on the number of electric appliances used, on the type of industry and on a variety of other factors that affect the consumption of energy.

To reduce the energy consumption in our daily lives it is not sufficient to use less energy, because a tremendous amount of energy still goes into manufacturing the equipment used. For example, car pooling would reduce gasoline consumption, but it would not reduce the energy that originally went into manufacturing the car. Enormous amounts of energy go into making the steel and the plastic out of which cars are made. It takes energy to operate the production machines that stamp out and assemble cars and indeed huge amounts of energy to into manufacturing the manufacturing machines themselves. This is why it is important to design with materials that are less energy-consumptive. To manufacture equipment made of recycled aluminum requires less energy than using virgin aluminum. Accordingly, the energy-conscious consumer should not only be mindful of how much operating energy an appliance will use, but should also be aware of how much energy was required to manufacture it.

3.4 WHAT IS ENERGY?

Because energy is a much discussed topic having great importance in our lives, it should be asked—what is energy? Scientists have contrived a mathematical entity that remains constant during mechanical, thermal, electrical, magnetic, or chemical processes. This entity is called energy. To give a good description that would put flesh on such an abstract concept is not easily done. In the case of energy one feels that there must be something physical to describe such a well-known entity; energy might be thought of as the potential for or presence of activity. According to Webster's dictionary it is the capacity to do work.

But what is work? In the dictionary there are numerous definitions of work. However, for the purposes of this chapter, *work* is done when a force moves an object across a distance. Thus, we can say that work is force times distance.

Machines, animals, and humans do work. In other words there is animate and inanimate energy. The more work an animal or machine does the more energy it uses. A large car can carry more passengers and do more work than a small car. Thus it uses more energy. But of course a train can carry even more passengers and freight than a car. Hence, it uses much more energy. Howver, what is economically important in transportation is how much energy is used per person per mile or per ton per mile. In the case of a train this number is much smaller than for a car. If we are to conserve energy and be economical, the amount of "work" should be small.

There is still another term, power, that is needed to conduct an intelligent conversation in energy-related subjects. By definition, *power* is the time rate of doing work; that is, how much work is done per unit time. For example, a motorbike can carry a passenger much farther in an hour than a horse or camel can, so the motorbike has more power.

Mankind has been using energy since the dawn of civilization. However, throughout the ages there have occurred changes in the form of energy used. Very early it was natural energy such as solar energy from the sun. Man then used animate energy by putting animals such as horses and buffalos to work. More recently, during the Industrial Revolution, machines and electrical energy were developed, and their use is increasing the world over.

In Figure 3.4 may be seen the decrease of animate energy and the increase of inanimate energy as industrialization has grown. It is unlikely that these trends will be reversed because, although they are more

FIGURE 3.4

COMPARISON OF CHANGES
IN ANIMATE AND INANIMATE ENERGY

Energy resource demands of food production

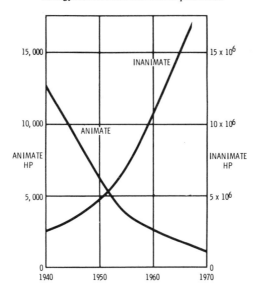

energy-consuming, inanimate-energy devices are more productive and convenient. Also, to substitute animals for machinery would take large amounts of grazing land that can be put to better use, such as the growing of food for people.

Energy resources have also changed over the years. Fuel wood has given place to coal, and coal, in turn, to oil and gas (see Figure 3.5). These trends may be expected to change, however. Research in the biomass area may once again give rise to the use of fuel wood, and already this is increasing in New England. With the rapid depletion of oil and gas, coal usage is increasing and may be expected to double in the next decade or so.

In summary, then, there is no one fuel which is the best for all times. It all depends on availability and societal demands such as convenience and environmental cleanliness. Clearly, there are trade-offs. For example, electric power plants used to burn coal almost exclusively. However, because natural gas was cheap, convenient, and environmentally benign, many power plants switched to natural gas. Now that natural gas is not abundant, power plants are being switched back to coal, of which we have an abundant supply. The burning of wood and biomass as a fuel may be increasingly used. In other words, the popularity of a certain fuel depends on the circumstances and may vary from time to time. Such changes will become more pronounced as petroleum and natural gas are depleted and new energy technologies come to the forefront. The changes should not scare us. As is evident from Figure 3.5, we have experienced fuel changes in the past and will continue to do so in the future.

FIGURE 3.5

U.S. ENERGY CONSUMPTION PATTERNS, 1850–1974

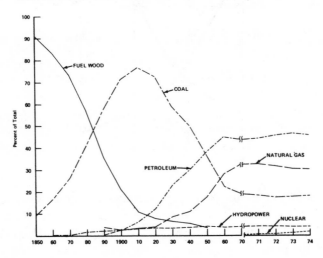

SOURCE: U.S Bureau of the Census; U.S. Bureau of Mines.

Units and Measurements

When we make a budget or balance our checkbook, we commonly specify the number of dollars of income and expense. Similarly, when we do energy accounting, we must specify the units or measures of the energy being accounted. Unfortunately, there is no one set of commonly used units in energy accounting. This gives rise to considerable confusion and comparisons are difficult to make. The problem is compounded because the United States is in a transition from the British System of units to the metric system. Scientists have switched from the CGS (centimeters-grams-seconds) System to the Système Internationale (SI).

The matter of units and measurements has been the subject of entire volumes. Clearly, that much detail is beyond the scope of this chapter. Accordingly, we summarize here the most common units and dimensions for force, energy, and power.

Force

$$1 \text{ Newton} = \frac{1 \text{ kilogram} \times 1 \text{ meter}}{\text{second}^2}$$

1 pound of force $= 4.448$ Newtons

Energy

$$1 \text{ Joule} = \frac{1 \text{ kilogram} \times 1 \text{ meter}^2}{1 \text{ second}^2}$$

1 Btu $= 778$ feet $\times$ 1 pound of force $\times$ 252 calories
1 calorie $= 4.186$ Joules
1 kilocalorie $= 1,000$ calories
1 kilowatt hour $= 3,413$ Btu

Power

$$1 \text{ kilowatt} = 1,000 \text{ watts} = \frac{3,413 \text{ Btu}}{\text{hour}}$$

$$1 \text{ watt} = \frac{1 \text{ kilogram} \times 1 \text{ meter}^2}{1 \text{ second}^3} = \frac{1 \text{ Joule}}{1 \text{ second}}$$

$$1 \text{ horsepower} = \frac{550 \text{ feet} \times 1 \text{ pound of force}}{1 \text{ second}} = \frac{2,545 \text{ Btu}}{1 \text{ hour}}$$
$$= 746 \text{ watts}$$

Approximate Energy Equivalents

1 cubic foot of natural gas $= 252$ kilocalories
1 pound of bituminous coal $= 2,898$ kilocalories
1 barrel of crude oil $= 1.46 \times 10^6$ kilocalories
1 gallon of gasoline $= 36,225$ kilocalories

There is still another set of multipliers, namely in terms of 10^n. Care must be exercised in using these because in the U.S. and in England different names are used for the same value of the exponent "n." Here we present the most commonly used exponents in energy accounting in the U.S.:

10^9 = billion (also called "giga")
10^{15} = quadrillion (abbreviated "quad")
10^{18} = quintillion (abbreviated "Q")

Thus, for example, the present U.S. consumption of energy (in Btu's) is about 73 quads or 0.073 Q. In other words, when somebody states that the U.S. consumes 73 quads annually he or she means 73×10^{15} Btu per year. Clearly this could have been 73,000 trillion Btu per year. All three are the same. However, what is wanted is to choose that "n" in 10^n that will give a minimum of digits and/or no decimal fractions. This is why generally when we give the amount of energy consumption we say 73 quads. In this way quintillions or other long numbers are more easily expressed.

3.5 FORMS OF ENERGY

Energy has many forms and different professionals classify energy differently.

Basically we have two types of energy resources: nonrenewable and renewable. The nonrenewable resources are the fossil fuels such as coal, petroleum, shale, and natural gas, and the fissile fuels such as uranium. The renewable resources are solar energy (including wind, ocean thermal gradients, and biomass), tidal energy, and hydro energy.

There are two important forms of energy that fit into neither category because although they are of long duration they are not, theoretically, renewable. These are geothermal and fusion energy.

Renewable resources constitute an ever-replenishable source of energy and may be expected to last as long as the solar system lasts. The nonrenewable resources were formed many thousands of years ago and cannot be renewed readily. All of our fossil and fissile fuels fall into this category.

Another way of classifying energy is by the physical form which it may assume. There are four physical forms that energy may take: (a) mechanical, (b) electromagnetic, (c) thermal, and (d) chemical/nuclear. Mechanical energy is the energy in moving bodies. Electromagnetic energy includes energy of such radiation as radio waves, visible light, and gamma rays, and energy in electric currents. Thermal energy is heat. Chemical or nuclear energy is energy residing in a molecule or atom which can be released by involving the molecule or atom in a chemical or nuclear process.

Energy utilization frequently requires conversion of energy from one physical form to another (some conversion processes are shown in Figure 3.6, where electromagnetic energy is divided into radiant and electricity forms). For example, when fuel oil is burned the combustion process is a chemical process in which the chemical energy residing in the fuel oil molecules is released as thermal energy. This thermal energy can be used to convert water to steam which, in turn, can be used to drive a turbine, thus converting thermal energy into mechanical energy. A generator, connected to the turbine, would convert mechanical energy into electrical energy. And passing electricity through a light bulb would convert electric energy into radiant energy.

These processes are examples of the interconvertibility of energy. If energy is quantified then the amount of energy in its various physical forms going into such a conversion process is always equal to the amount of energy in its new physical forms which come out of the process. This

FIGURE 3.6

ENERGY-CONVERSION PATHWAYS

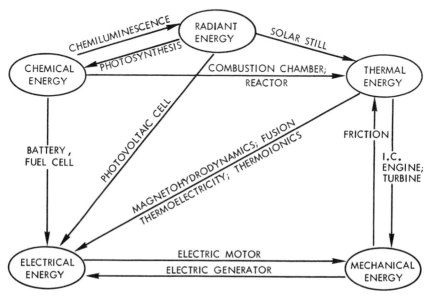

fact is called the First Law of Thermodynamics and is a quantification of the definition of energy given earlier. More crudely, it says energy is neither created not destroyed in a thermodynamic process.

3.6 THERMAL EFFICIENCY

Consider the automobile. Its gas tank contains gasoline which is a repository of chemical energy. The gasoline is mixed with air and, when ignited, releases thermal energy. Not all of this thermal energy is converted into mechanical energy in the drive shaft; some of it is lost in the form of hot gases through the exhaust pipes. No matter how the automobile is designed, the thermal energy cannot be entirely converted to mechanical energy; some thermal energy always remains. The First Law of Thermodynamics is not violated by this inability to convert all of the thermal energy into mechanical energy:

Heat Supplied = Useful Work + Heat Rejected + Frictional Losses

The problem is that heat does not exist apart from physical material and losing heat causes the temperature of the material to drop. The temperature of the material will not fall below the temperature of its ambience.

Schematically, the process whereby heat is transformed into useful work is shown in Figure 3.7. The hot reservoir supplies thermal energy Q_H at temperature T_H, and the cold reservoir receives thermal energy Q_L at temperature T_L. The difference

$$W = Q_H - Q_L$$

is the useful work done.

Thermal efficiency is defined as

$$e = \frac{useful\ work}{supplied\ heat} = \frac{Q_L}{Q_H} = \frac{Q_H - Q_L}{Q_H} = 1 - \frac{W}{Q_H}$$

In an idealized model this can be rewritten

$$e = 1 - \frac{T_L}{T_H}$$

FIGURE 3.7

SCHEMATIC DIAGRAM OF A REVERSED-HEAT ENGINE

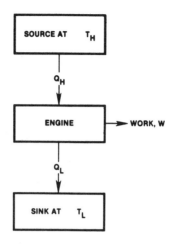

where T_L and T_H are the absolute temperature* of the cold and hot reservoirs respectively. For this idealized process to be 100 percent efficient either the exhaust temperature T_L must be absolute zero or the supply temperature T_H must be infinite. Since neither of these possibilities is met in practice, thermal efficiency, that is, the percentage of thermal energy converted to useful work, is always less than 100 percent. Moreover, this idealized model is just that, ideal. Actual processes with some T_H, T_L never work as efficiently as the ideal model operating with the same T_H, T_L. In Figure 3.8 it can be seen that about half of the energy·is lost.

Thermal efficiency should be understood as an important feature of one of the central facts in the modern "problem" of energy: the vast bulk of the energy consumed in the United States comes from the consumption of fuel in order to release thermal energy which will then be converted to some desired form. The engineer who studies these conversion processes will be alert to possibilities for improving the efficiency of the conversion process, not only for economical reasons but also because the exhausted thermal energy is not necessarily environtally benign. A nuclear power plant, for example, typically uses water in

---

·Kelvin = Celsius + 273, Rankine = Fahrenheit + 460 are absolute temperature scales.

FIGURE 3.8

EFFICIENCY OF ENERGY UTILIZATION, 1960–1985
(Business as usual; $11 oil)

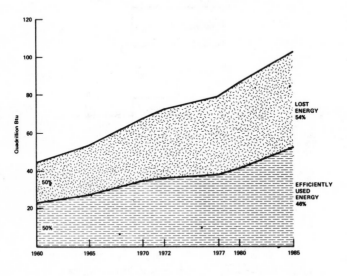

SOURCE: U.S. Department of the Interior.

at least part of the process of transferring thermal energy from the reactor core to the electricity generating equipment. The unused thermal
energy is unleashed on the local environment by the power plant in the
form of water which is warmer than the ambient water. This warmer
water can have an ecological impact. It is in the self-interest of the power
company to remove part of the basis of local resistance to the construction of power plants by finding ways to eliminate this ecological impact.

3.7 ENERGY SUPPLY AND DEMAND

In our daily lives all of us must live within our budget, i.e., our capital
and income on the one hand and our expenditures on the other. So it is
with energy budgeting. We have a capital in nonrenewable resources
such as coal, petroleum, gas, and uranium that we can tap. We have an
income from renewable sources such as solar energy that is comparable

to a steady paycheck. On the other hand, we have ever increasing expenditures due to the increasing population and greater affluence in our life styles. We must carefully budget our energy resources and expenditures. In other words we must be aware of energy supply and demand. In doing so we can not merely balance supply and demand for at least two reasons: (1) we must think of future generations—it would be unfair to leave them in a world exhausted of energy resources; (2) not all energy resources are used to provide energy but some resources are used as feedstock in manufacturing such as petrochemicals in the case of plastic products.

When we speak of energy resources we need to consider different implications. From a physical resource point of view we arrive at different conclusions than we do from economic or political standpoints. There is no energy resource shortage and there never need be one as long as the solar system exists. However this is not true socially, politically, or economically. Politically, we may have embargos which could interfere with our technological development. Economically, the price of energy may go up so high that some people are priced out of the market. This paradox of unlimited energy resources at the same time we have an energy shortage can only be reconciled by prudent energy management and modern technology.

The demand for energy is commonly given by end use sectors (see Figure 3.9). These are: (*a*) household and commercial (where commercial represents office buildings, hospitals, etc.); (*b*) industrial; (*c*) transportation; and (*d*) electrical. Of course not all end users rely on the same energy resource and, as we have seen, there has been a constant shift from one source to another and this may be expected to occur again. For example, if the research and development being conducted in biomass is successful, fuel wood may become important again. The shortge of petroleum and natural gas may bring about a shift to coal, uranium, solar, and geothermal energy.

In evaluating the increasing energy consumption, different sources use different figures. This makes any comparison difficult. There are three main reasons for the confusion.

First, some people speak of the total energy consumed while others talk of only the increase in electrical energy. It is projected that total energy use will increase at the rate of approximately 3 percent per year while use of electrical energy will increase at the rate of about 6 percent. Exactly what these rates of increase will be depends on how successful we are in enforcing conservation measures.

FIGURE 3.9

U.S. GROSS ENERGY CONSUMPTION BY SECTOR, 1960–1973

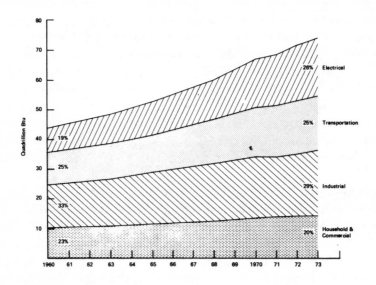

SOURCE: U.S. Bureau of Mines.

Second, different authors use different periods in time when extrapolating historically. Energy use does not increase monotonically; rather, the rate of increase fluctuates. After years of increasing, the total gross energy consumption declined by 2 percent in 1974. Thus it is important to realize the assumptions that are made when making future projections.

The third problem, which is related to the first, is that not all four sectors behave similarly. The rate of increase in the household and commercial sector depends on structures being built, while the consumption by the industrial sector depends on new capital investment as well as on plants that are idle. Transportation depends on traffic, size, and age of vehicles, and the electrical sector depends on power plants brought on-line.

Clearly energy projections must be made with great care and they must be interpreted with careful scrutiny.

3.8 ENERGY SYSTEMS

As may be seen in Figure 3.10, energy systems may be divided into proven, promising, proposed, and speculative systems.

The proven energy systems are those with which we have considerable experience. Primarily they include coal, petroleum, and gas-burning power plants as well as nuclear and hydroelectric facilities. Each has its problems. For example, low-sulfur coal does not exist in abundance everywhere and hence many coal-burning power plants have the alternatives of emitting sulfur oxides and being in violation of EPA requirements or installing expensive emission control equipment which adds to the already high cost of power plants. The result is that the increased cost of electricity is passed on to the consumer.

Petroleum and natural gas are relatively clean, environmentally speaking, but they are scarce in supply.

Nuclear energy is an energy system beset by controversy. If the cost of building nuclear power plants can be kept under control, nuclear energy can be a source of low-cost electricity for the future. The real battle over nuclear energy, however, is being fought over the safety and safeguard issue: how safe are the nuclear plants and how well is the nuclear material safeguarded? Radiation from a reactor plant operating normally is less than naturally occurring radiation. The problem is that an accident such as an unexpectedly severe earthquake, a failure such as a fatigue failure in the main coolant piping, careless operation, or even sabotage, however implausible any of these events, could result in a serious incident in which dangerous amonts of radioactive material escape from the containment system. Moreover, more nuclear plants mean more nuclear fuel being manufactured, shipped, and stored. It is not inconceivable that a terrorist group could acquire some and build a bomb. Finally, the disposal of radioactive waste products of nuclear plants is becoming a mounting problem.

Hydroelectric plants are environmentally benign and use a renewable energy resource, namely the potential energy of stored water. In general their upkeep is low although the capital cost is high. Further, they are localized and can be constructed only in favorable locations.

Geothermal and tidal energy can produce energy but they also are both localized. In general, they are both environmentally benign and fuel costs are nonexistent although capital costs may be appreciable. Italy relies heavily on geothermal energy and approximately 5 percent of the electricity of San Francisco is supplied from geothermal energy.

FIGURE 3.10
ALTERNATIVE ENERGY SYSTEMS

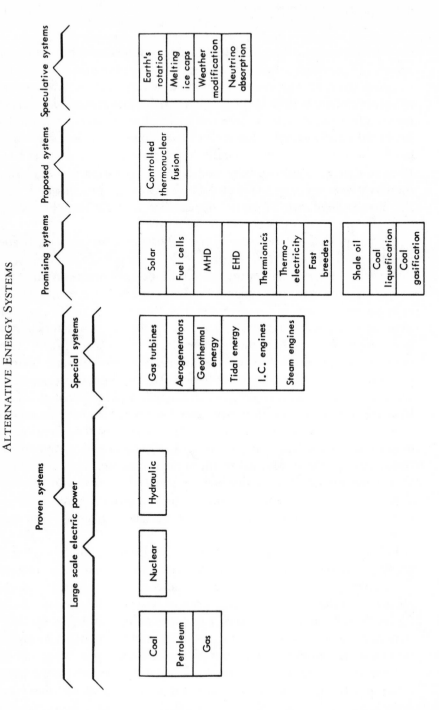

Tidal energy, another localized energy source, is relatively little exploited. It is used in France and in the Soviet Union.

Aerogenerators or wind energy date back to ancient times, when they were used primarily to pump water or to produce the mechanical energy to grind flour. More recently, they are being considered for generation of electricity. Although they are relatively benign environmentally and have no fuel costs associated with them, the wind does not blow steadily; hence, as producers of electricity, they require a storage system such as batteries. However, batteries deliver direct current whereas most appliances are built for alternating current. Thus there is a mismatch between the delivery system and the apparatus to be used.

Internal combustion engines are widely used in automobiles, lawn mowers, trucks, and tractors. They are highly reliable and relatively cheap. However, they consume considerable energy and are not environmentally benign. They come in two general forms: (1) the spark ignition engine (Otto Cycle), most commonly used in passenger vehicles, and (2) the compression ignition engine (Diesel Cycle), most commonly used in trucks and tractors.

Stationary gas turbines, a spinoff from aircraft gas turbines, are now used to provide auxiliary power.

Solar energy has received much attention in recent years. There are many reasons: (a) Solar energy is abundant. If only 2 percent of the land mass of the U.S. were to be used as a solar collector and if the conversion efficiency to useful energy were to be 10 percent, solar energy alone would meet all of our energy needs and then some. (b) Solar energy is a renewable resource and will not be depleted as long as the solar system exists. (c) Solar energy is environmentally benign. (d) With solar energy there is no fuel cost.

Some of the ways solar energy may be exploited are: (1) It may be used for heating and cooling of buildings. (2) It may be used for heating water. (3) Electricity may be generated from solar energy by means of photovoltaic cells or films. (4) Electricity may be generated in a solar thermal power plant. (5) Wind energy may be created by thermal gradients. (6) Ocean thermal energy conversion (thermal gradients in the oceans with warmer water on top and cooler water deep down) can be exploited to operate thermodynamic cycles. (7) The sun provides energy for photosynthesis by means of which biomass grows. This biomass can be converted into other fuels or it may be burned.

Of the above ways, solar heating and cooling of buildings by flat plate collectors and concentrators has been discussed most commonly and is gradually gaining acceptance. However, it does have certain disadvantages. Solar radiation is intermittent, hence the energy must be stored for the times when the sun is not shining. At present, market rates for solar installations are relatively high. This is exacerbated by the fact that, in general, a solar system must be accompanied by an auxiliary system. The fact that a solar system may require a higher capital investment but does provide a relatively low maintenance cost does not impress the average home buyer because the average ownership of a house in the United States is seven years whereas a solar heating installation pays for itself in about fifteen years. The higher principal and mortgage rates discourage homeowners from installing solar heating and cooling systems. The solution lies in institutional arrangements such as tax benefits and government subsidies and loans for solar installations. Even if this were the case it would take many years before solar energy would occupy a major share of the heating and cooling of buildings. Every year there are about 1.5 million housing units built in the U.S., whereas there are in existence about 90 million housing units. Thus even if every new housing unit were heated and/or cooled by solar energy, the difference would be minuscule for many years.

But flat plate collectors and concentrators are not the only direct ways of using solar energy. A promising way of exploiting solar energy is by means of photovoltaics, namely films or wafers that produce electricity when they are irradiated. The most common photovoltaic cell may be found in your camera's lightmeter which tells the amount of light incident on the photographed object. Photovoltaics was also very successfully used in the space missions. The main problem in widespread commercial use is cost. Still another form of solar energy utilization is solar thermal power. In this case a concave mirror is mounted on a hollow tower. The tower is surrounded by heliostats which focus the reflection of the sun on the mirror, which in turn focuses it on a steam boiler that operates a steam turbine and rotates an electric generator.

The exploitation of solar energy may be slow in coming because, at the present time, it is not cost effective. Although not imminent, it could be indispensable in the years to come. The production of petroleum and natural gas will peak in 1990–2000. In the meantime, we will be using more and more coal for burning, liquefaction, and gasification, and our presently abundant coal supplies will ultimately be depleted. Still, using more coal will borrow us time. The other source of energy that we will probably use more of is uranium. However, our supply of high-grade uranium ore is not plentiful and the problems of nuclear waste management are not solved. As technological breakthroughs occur the cost of solar energy will go down and thus solar energy will gradually come into its own. When this happens, the fear of an energy catastrophe will recede.

Another promising energy source is the application of controlled thermonuclear fusion. Controlled thermonuclear fusion is the opposite of fission in which heavy elements are broken down. In fusion, light elements are brought together, releasing tremendous amounts of energy. The light element to be used in fusion could be deuterium, which is plentiful in the seas. Thus, there is a huge supply of free energy. Unfortunately, we still don't have an operable, controlled fusion reactor. Presently, researches are working on two types of controlled thermonuclear fusion: (1) magnetic confinement of the plasma (i.e., ionized gases); and (2) laser fusion. It should be noted, however, that the commercial, practical use of controlled nuclear fusion is many decades away.

Finally, there is magnetohydrodynamic (MHD) power generation. Here a hot ionized gas is passed between the poles of a magnetic field, thereby generating electricity. The high efficiencies of MHD generators would save fuel and cooling water.

One may be sure that with the vast amounts of money and economic power that are at risk as an energy industry emerges from a reasonably stable past and prepares to meet an uncertain future, decisions will not be made just on technological grounds. Energy consumers—that's us—should be conscious of their self-interests and work to make governments and businesses responsive to those interests. The energy systems that will be set up to replace the petroleum system will have to be so large and their organization will cost so much in investment and in time that their implications should be understood and debated by an enlightened public. Moreover, besides tax and regulatory policies designed to promote one or another alternative energy system, the federal government can encourage universities to develop a corps of energy engineers. An energy engineer would on the basis of technological analyses and economic projections be able to design or advise in the modification of a complete energy system for a house, a neighborhood, a factory, a city, or a region.

3.9 EXERCISES

1. Write a short essay on how to educate the public to be more energy-conscious.

2. How do you interpret the present energy crisis? What would be the remedies?

3. Study Figure 3.2 and discuss the reasons behind the indicated positions of a few countries, i.e., Switzerland vs. USA, etc.

4. Do you foresee any drastic changes in the near future in the U.S. energy consumption pattern shown in Figure 3.5? Discuss.

5. Can you name a few household appliances where the energy is very inefficiently utilized? Discuss the reasons.

6. Cite a few energy systems and discuss their advantages and disadvantages.

7. Write a short essay on the sources of energy that most environmentalists recommend. They are often referred to as "clean energy."

3.10 BIBLIOGRAPHY

1. Cambel, Ali B., and Richard C. Warder. "Energy Resource Demands of Food Production." *Energy*. vol. 1, 1976, pp. 133–42.
2. Dresner, Stephen. *Units of Measurement*. New York: Hastings House, 1971.
3. Enzer, Hermann; Walter Dupree; and Stanley Miller. *Energy Perspectives*. Washington, D.C.: U.S. Department of the Interior, U.S. Government Printing Office, 1975.
4. The Ford Foundation. *Exploring Energy Choices—Energy Policy Project of the Ford Foundation*. 1974.
5. O'Toole, James, and the University of Southern California Center for Futures Research. *Energy and Social Change*. Cambridge, Mass.: MIT Press, 1976.
6. Saperstein, Alvin M. *Physics: Energy in the Environment*. Boston: Little, Brown, 1975.

Chapter Four

TRANSPORTATION AND TECHNOLOGY ASSESSMENT

4.1 CHOICES IN PASSENGER TRANSPORT

> The delight which comes of rapid movement has never been under-
> stood until one occupies a place in a horseless carriage on a smooth
> road. There is an exhilaration from the swift motion surpassing that of
> any other form of movement. The Stanley Carriage is capable of a speed
> from 30 to 42 miles per hour,* according to the gear used, and racers
> are made to exceed even this.
>
> *—1899 ad for Stanley Steamers*

Ever since people have attempted to travel from one place to another,
the primary criterion for selecting a mode of transportation has been the
speed of the vehicle. Criteria such as safety, privacy, and comfort were
important but in the final analysis speed was valued most. For some
reason, we believe that time is money. Before the 1960s, faster was
almost always better.

In the past twenty years, some of the negative consequences of increasing the speed of vehicles have caused many people to question its
value. The 1973 shortage of gasoline brought about a national 55-miles-
per-hour speed limit on our highways, which were originally designed
for higher speeds. A fringe benefit has been a decrease in automobile
accidents.† The Supersonic Transport (SST), designed to halve the time

---

*In the metric system, the Stanley Carriage would be capable of speeds from 48 to 67
kilometers per hour.

†Since the law took effect, there has been an approximately 20 percent reduction in
highway fatalities.

of transoceanic flights, has met with great public opposition due mainly to environmental and economic concerns.

Speed is only one facet of the problems of passenger transport. A more significant issue is the importance of reversing the trend of decreasing usage of public transit. This trend is the result of millions of individual choices; Table 4.1 provides information about choices made in the New York metropolitan area (2).

TABLE 4.1

Auto versus Public Transit Trips in New York Region

| Year | Estimated trips on an average weekday, in millions | | |
	Transit	Auto	Total
1950	15	15	30
1970	10	30	40
2000 (based on current trends)	8	60	68

In considering the ways of increasing ridership on public mass transit systems, factors such as quality and cost of the available systems are of great importance. However, since these systems are in competition with the private automobile, its influences on the public need to be incorporated in transit planning. This example is presented to illustrate the need for a comprehensive approach to decision-making about sociotechnological problems and issues.

Technology Assessment is the name for one such approach. When considering alternatives to the automobile, both its advantages (benefits) and disadvantages (costs and risks) need to be considered. An assessment team in studying this question might start by coming up with the information presented in Table 4.2.

In the eyes of the auto user, the advantages appear to outweigh disadvantages. This is due in part to other-than-transportation reasons for owning a car. An impact analysis would show that social status, need for privacy and freedom, and pride of ownership are some non-transportation reasons for choosing cars.

A second reason for an individual's perception that the benefits outweigh the costs is related to the idea of the "Tragedy of the Commons." All of the advantages benefit the individual directly while many of the disadvantages cost society as a whole in terms of increased pollution, congestion, and inefficient use of resources. Therefore, the individual

TABLE 4.2

ADVANTAGES AND DISADVANTAGES OF AUTOMOBILES

Benefits	Costs and risks
Good acceleration and high top speed	Automobile accidents are a major contributor to injuries and deaths
Origin-to-destination system	Land is being used up for more highways and streets
Available at all times	Cars need to be stored in parking spaces, garages, etc.
Personal transport—provides privacy and comfort	Internal combusion engine is very inefficient. Wastes fuels and construction materials
Easy to handle—can go in any direction	Expensive to buy, operate, and repair. One of the worst investments that people can make
Cargo space is available	Major contributor to air pollution and adds to solid waste problem

will continue to use the automobile because he is not held directly responsible for adverse effects on the "commons," which is our air, land, and other resources.

A very promising solution to our urban transportation problems may well lie in developing a concept of mass transit that offers the unique advantages of the automobile and avoids many of the disadvantages. In terms of attracting the public to mass transit, we would do well to take as a starting point the most attractive personal features that drew the American to the automobile.

It is interesting to note that in the beginning of the twentieth century, the automobile was being hailed as the solution to one of the worst pollution problems of our cities. The waste products from horses were beginning to choke our cities and the "clean" automobile was certainly going to improve the environment. This is another illustration of the need for technology assessment, which includes technological forecasting.

The remaining sections of this chapter will provide specific examples of the assessment of technologies that relate to passenger transit. To provide a foundation for discussion, the what, why, who, and how of technology assessment (TA) are first described. Then specific case studies and how TA is applied to transportation problems and technologies are discussed. The chapter concludes with a discussion of technological forecasting and future transport systems.

To conclude this section, a historical illustration of the fact that alternative technologies can come in many ways is described. In 1902 Ignaz Schroppe invented a horsemobile which was designed to overcome many of the perceived disadvantages of the automobile. Pictured here is Schroppe's invention, which he designed for the following reasons:

> The automobile needs a relatively smooth and wide thoroughfare; it does not drive over stones, cannot cross fields and forests. Its main flaw lies in the wheels. Man was endowed with legs, not wheels. That is why I invented an automobile with legs. The engine sits inside the body and controls each of the four legs. Exhaust fumes escape via the most likely spot of the construction. The steering wheel controls each pair of legs separately. Collapsible trays on each side of the body may be used by the driver for eating or playing cards. (11).

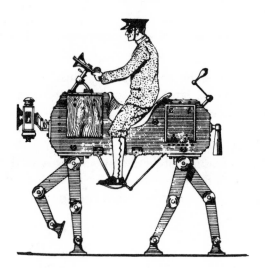

4.2 TECHNOLOGY ASSESSMENT

In order to make wiser decisions concerning the use of technology, the impacts of alternative courses of actions as well as their relative benefits and costs need to be studied. The systematic study of the effects on society that may occur when a technology is introduced, extended, or modified with special emphasis on the impacts that are unintended, indirect, or delayed is called technology assessment (TA). TA emphasizes the secondary effects of a new technology and attempts to provide a balanced look at all the alternatives, options, and possible outcomes (6).

People have always been interested in the consequences of technological development. One might reasonably ask, therefore, what's new about TA? Several things are new. First is the attempt to expand the study of the anticipated consequences well beyond the conventional considerations of economic costs and benefits or immediate implications for the perpetrator or user of technology. The extension into the full range of social, economic, political, environmental, international, legal, and other impacts demands a degree of skill and sophistication rarely brought to the management of technology. The second feature that is unique to TA is its focus as a policy study on informing the interested publics and decision-makers of the possible range of consequences for new actions.

Historical Background

This modern comprehensive approach to assessing the consequences of using technology had its beginnings in the mid-1960s. The concern for the environmental impacts of our technological systems had finally reached the general public. In 1969 the Environmental Policy Act (Public Law 91-190) was passed by the Congress, and it required that an environmental impact statement be filed by every U.S. government agency that was planning a project. Assessing environmental impacts is only one facet of TA, but it was a beginning.

Some members of Congress recognized that in order for them to make better-informed decisions, technology assessments on technologically based issues and problems were needed. After much debate the U.S. Congress established an Office of Technology Assessment (OTA) which was signed into law on October 13, 1972.*

Other federal agencies such as the U.S. Patent Office and the Library of Congress have also become involved in TA-type studies. Some re-

---

*An annual report is submitted by OTA to Congress. Copies can be purchased ($1.55) by writing to Superintendent of Documents, U.S. Government Printing Office, Washington, DC 20402.

gional and state planning groups have also begun to work on limited assessments of technology. One source of funding for TA studies is the National Science Foundation. In the 1970s private study groups (think tanks) such as Mitre Corporation and Arthur D. Little, Inc. had turned TA into a business. Their clients include private companies as well as government agencies who want to anticipate the potential and impacts of new technologies.

Is TA Needed?

The funding and time needed to carry out technology assessments vary greatly depending on the scope of the studies. A moderate-size study might involve a team of twenty people working for a year. A study of this magnitude could cost from two to three hundred thousand dollars. Is the expenditure of human and monetary resources justified? Are TA's really needed? Will results of assessments contribute to wiser decisions?

Several trends make it virtually mandatory that society develop new warning techniques and better aids to planning and decision-making. First, the growing complexity of society has increased the magnitude and significance of secondary impacts of technology. In our highly interwoven and interdependent society, disorder in a single component can create havoc throughout society, as in 1965 and 1977 when power failures blacked out New York City, causing major dislocations in all sectors of life.

The larger scale of human enterprise and our increasing power over nature is another trend that demonstrates the need for TA. Building of oil pipelines, weather modification, and construction of superhighway systems are conducted on such an enormous scale that they demand gigantic investments, involve long planning and implementation periods, and engender intractable, if not irreversible, consequences. New technologies such as the SST, nuclear breeder reactors, and recombinant DNA techniques represent a giant step in our ability to travel faster, generate more energy, and better understand and control genetic-related problems. However, they also have severe negative consequences which have to be weighed against the benefits before decisions concerning their future development and usage are made.

As the science and technological establishments grow, almost every aspect of our daily lives is affected by technology. Therefore, in order to satisfy human needs and find solutions to societal problems, both the capabilities and limitations of technology must be clearly understood. TA is needed to provide decision-makers with information so that optimum choices can be made. The findings of a TA study should be presented in a manner that will facilitate follow-up action such as:

—Modification of proposed project.
—Surveillance of chosen system.
—Stimulation of new research and develoment.
—Establishment of controls.
—Passage of new laws that may be required.
—Blocking of a project.
—Selection among alternative proposals based on trade-offs.

TA Methods

Since the first formal TA studies were being conducted only approximately ten years ago, study groups are still experimenting with different methods and techniques. However, in the past ten years, certain patterns have emerged that provide some guidance in conducting TA's. The first decision of a study group is to decide on the focus of their TA.

Technology assessments can be categorized as being of two types:

Problem-Initiated Asssessment (PIA)—This type of TA involves the study of alternative solutions to a specific sociotechnological problem such as the energy crisis or air pollution caused by the automobile.

Technology-Initiated Assessment (TIA)—This type of TA involves the study of the potential uses and impacts of an emerging or futuristic technological system such as weather modification or the SST.

This initial separation may seem trivial, but in fact helps specify study goals and provides a basis for limiting the scope of the study. Since human resources and funding are always limited, careful planning must go into the selection of areas to be studied in depth. Table 4.3 outlines procedures that have been used successfully by small study groups (two to five people) for conducting a limited TA.

The next two sections of the chapter will provide specific examples to show how TA concepts can be applied to passenger transit technologies and problems. PIA studies of ways to decrease energy intensiveness of passenger transport systems and of proposals for improving ferry service across the Long Island Sound will be discussed. Next, a summary of a TIA study on the Concorde SST is presented. Technological forecasting, an integral part of TA studies, will be discussed in the fifth section. The chapter concludes with a look at the future of passenger transit.

TABLE 4.3

Procedures for Doing a Limited TA

Since study groups have only a limited amount of time to do their assessments, it would be impossible to carry out a comprehensive TA. Therefore, the following procedures are suggested for narrowing down the scope of a group's TA. First, decide on the type of TA project—choose between PIA and TIA. If you choose PIA, use Procedure A; TIA, use Procedure B.

Procedure A—Outline for Doing PIA Assessments

1. Describe the sociotechnological problem to be assessed. This section should include analysis of magnitude of problem, and people and/or environments affected by the problem.
2. Develop a decision tree for the identification of alternative policy options for ameliorating the problem. This analysis should include categorization of options such as new technology, education, and legislation.
3. Delineate the scope of the assessment task. This decision is constrained by the time available and the number of people in the assessment team. Select as many options for comparison as is appropriate.
4. Describe the background information of each option that is being assessed.
5. Analyze the positive (benefits) and negative (costs and risks) aspects of each option. This would include an impact analysis of each option on affected systems and people.
6. Compare the pros and cons of alternative policy options.
7. Recommend policy options for implementation and preferred ways of implementing them.

Procedure B—Outline for Doing TIA Assessments

1. Describe the emerging or futuristic technological system under assessment. This section should include technological forecasting and hardware and software aspects of the technology being studied.
2. Identify the major impacts of the technology on other systems and people. This section should include preliminary identification of impacts in terms of benefits and costs.
3. Delineate the scope of the assessment task. This decision is constrained by the time available and the number of people in the assessment team. Select as many positive and negative impacts for study as is appropriate. Naturally, the more significant impacts should be selected first.
4. Analysis of the magnitude and significance of the impacts chosen for study.
5. Comparison between the positive impacts (benefits) and the negative impacts (costs and risks) of the technology being assessed.
6. Recommend policy for future use of technology being assessed.
7. Suggest monitoring techniques, if necessary, and enumerate the variables that should be monitored.

4.3 PROBLEM-INITIATED ASSESSMENTS

March 7, 1967: Congressman Emilio Q. Daddario introduced H.R. 6698 "to provide a method for identifying, assessing, publicizing, and dealing with the implications and effects of applied research and technology" by establishing a Technology Assessment Board.

July 3, 1967: The National Academy of Engineering (NAE) is invited by Congressman Daddario to consider making a test of the TA concept.

February 9, 1968: In accepting the invitation, the NAE statement stressed the experimental nature of the project and emphasized that the activities being proposed should be viewed as first steps toward finding means to improve or complement the present means of assessment of technology by the Congress.

The sequence of events described above resulted in a report entitled "A Study of Technology Assessment" (10) which was published in 1969. The idea of categorizing assessments into two classes was first discussed in this study. Problem-initiated assessments were viewed as studies with well-defined goal(s) which dealt with finding solutions to sociotechnological problems, while technology-initiated assessments were more concerned with studying the consequences of specific emerging and futuristic technologies.

In discussing problem-initiated assessments, the authors of the NAE report emphasized that:

> Systems analysis has unquestionably been of considerable value to the planner. But its successes have been limited to problems that have one characteristic in common: almost all the variables are focused toward a well-defined goal, namely, the solution of a problem. Air pollution caused by automobiles, aircraft noise, degraded fresh water resources, insufficient low-cost urban housing and rising health care costs are examples of problems that have impact on society and that can be and, in most cases, are likely to be satisfactorily assessed by the cause-effect methodology. (16)

Two specific sociotechnological problems related to passenger transport are discussed in this section to show how the PIA method discussed in the previous section can be applied. First, the national problem of finding and implementing ways to increase transportation energy efficiency is described. Then the regional problem of comparing alternative plans for increasing ferry service across Long Island Sound is analyzed.

Energy Intensiveness of Alternative Transport Systems

One of the most pressing sociotechnological problems that faces the United States as well as the world is to find more efficient ways of using alternative energy sources to replace our fossil-fuel resources. As discussed in the previous chapter, oil and natural gas are the most limited of our energy resources. Since transportation systems use approximately 25 percent of our energy budget, an examination of transportation energy use is timely and important. At least five reasons can be cited for conducting an assessment of alternative ways of increasing the efficiency of transport systems.

Transportation in 1970 required 16.5 quadrillion (10^{15}) Btu's, equivalent to 3 billion barrels of oil. Between 1950 and 1970, energy consumption for transportation had an average annual growth rate of 3.2 percent, more than double the U.S. population growth rate. A second reason is that world oil reserves appear to be quite limited. One estimate reports that 90 percent of the world oil supply will be consumed by about 2030. Third, U.S. transportation is almost entirely dependent on oil as a fuel. Fourth, exploration, production, transportation, refining, and use of petroleum present serious environmental problems. Finally, transportation contributes to a number of other environmental problems including urban congestion, inefficient land use, and noise.

Once a sociotechnological problem has been described and its importance established, a description of alternative methods of achieving the assessment goal is needed. One way of depicting decision options is via the use of a decision mobile (14). Figure 4.1 outlines a set of energy policy alternatives that are available. This taxonomic approach to categorizing alternatives first divides all options into supply and demand subsets. Since we are interested in identifying, studying, and comparing alternative methods of increasing energy efficiency, the decision tree is limited to options that relate to this goal.

In order to decrease energy intensiveness,* three major categories of alternatives are available, namely: modal shifts, increased load factor, and technological changes. Significant energy savings are possible via modal shifts because of the large variations in energy intensiveness of

---

*Energy intensiveness (EI), the inverse of energy efficiency, is used in subsequent discussions because data are most readily available in this form. EI is defined as Btu per passenger-mile.

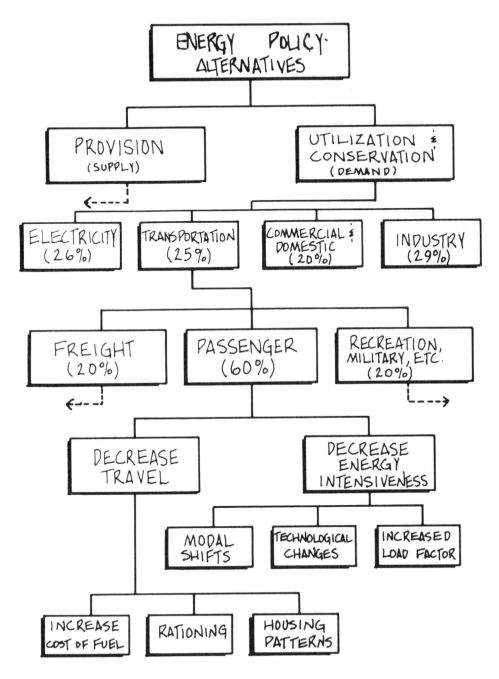

FIGURE 4.1

DECISION TREE OR MOBILE FOR ENERGY POLICY ALTERNATIVES

modes of passenger travel (see Figure 4.2). For example, in 1970 inter-city travel by auto was approximately twice as energy intensive as inter-city bus transport. Figure 4.2 also shows that from 1950 to 1970, except for railroad travel, the energy intensiveness of all of the other modes of travel increased significantly.

FIGURE 4.2

HISTORICAL VARIATION IN ENERGY INTENSIVENESS OF PASSENGER MODES

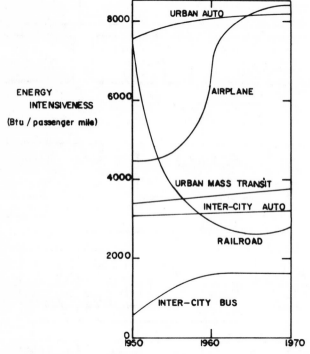

SOURCE: E. Hirst, "Transportation Energy Use and Conservation Potential" (8).

As outlined in Table 4.3, a problem-initiated assessment also studies costs, risks, and benefits of each alternative, which mainly involve primary impacts on individuals and society. Secondary impacts also need to be studied to guard against the implementation of a plan that may do more harm than good. In this assessment of alternatives for lowering energy intensiveness, implementation techniques such as economic incentives, public education, and government action also need to be assessed carefully before usage.

The example just discussed and other examples which will follow are included to provide the reader with an overview of the assessment process. At best, they provide summaries and outlines of assessments. To conclude this section, a second example is provided to illustrate the assessment of a regional transportation problem.

Improving Ferry Service across Long Island Sound

The need to improve the transportation system across Long Island Sound has increased as the population of Long Island (Nassau and Suffolk counties) has increased to over 2.5 million in 1975. * In the early 1970s, new proposed bridge crossings met with strong public opposition and were subsequently abandoned.† With bridge-building virtually eliminated and the transportation need still unmet, the New York and Connecticut Departments of Transportation decided to consider ways of improving ferry routes and services. A better ferry system would ease the difficulty of travel across the Sound at a fraction of the cost of a bridge.

The Tri-State Regional Planning Commission was selected to conduct the study. Its purpose was "to examine a variety of potential ferry routes and determine the costs and benefits of each route. This involved a careful analysis of ferry vessels, terminals, sites, and volumes of vehicles and passengers. Also evaluated was the impact that such a ferry service would have on the environment and local development" (3).

The basic choices that need to be made in setting up an improved ferry service‡ and the main consequences of these choices are shown in Figure 4.3. Nine crossing sites were considered. The study was designed to seek information about ferry performance, revenue projections, the cost of providing service, development impacts, environmental impacts, and to summarize costs and benefits.

A 56-page report discusses the findings, and interested readers can write to the Tri-State Planning Commission for a copy of the report (3). A summary of the conclusions will be presented next to provide the reader with the results of a study that hopefully will help transportation planners make wiser decisions. Some of the significant conclusions were:

Currently all traffic leaving Long Island must travel west and exit via New York City (except for very limited service on two ferry routes).

†For further discussion on the consent of the governed, refer to section 1.5.

‡Currently there are two ferry systems in operation.

FIGURE 4.3
FERRY SERVICE CHOICES AND CONSEQUENCES

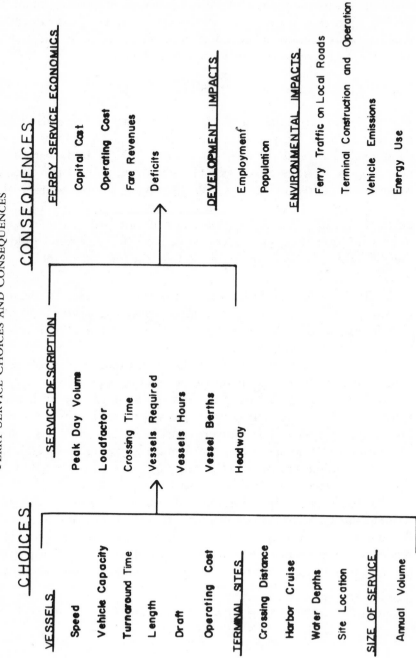

SOURCE: *Crossing the Sound* (3).

1. *Vessels:* A fleet of 200-vehicle-capacity displacement vessels would be more expensive to operate than smaller 100-vehicle-capacity vessels for volumes up to one million vehicles per year. The high-technology faster crafts are too expensive to purchase and operate in vehicle-carrying service.

2. *Size of the Ferry Service:* The ferry volume should be set at a point where the ferry fares plus the dollar value of external benefits (such as pollution and road use saved by the ferry) are equal to the marginal cost of providing ferry service.

3. *Development Impacts:* The heaviest impacts expected in any area are less-than-one-percent increases in eastern Suffolk population and employment as a result of a 500,000-vehicle-per-year ferry from Orient Point.

4. *Environmental Impacts:* At most sites the most significant impact of the ferry is the traffic that will be generated on local access roads. The peak-day, peak-hour traffic is expected to be as much as 40 percent of the capacity of single-lane traffic on a rural highway under relatively uncongested conditions. This represents additional traffic on local access roads.

A problem-initiated assessment should be designed to provide decision makers with objective information and rational recommendations. In the ferry study, both near-term and long-term recommendations were suggested.

4.4 THE CONCORDE SST—
A TECHNOLOGY-INITIATED ASSESSMENT

When the cause-effect method of assessment is applied to technology-initiated situations, the results will differ substantially from the results that can be obtained from problem-initiated assessments. While the latter focuses on solving a stated problem, the process involved in an assessment of a new technology is better represented by an analogy with an inverted funnel. The assessment process begins with the new technology at the small end and emerges as a complex pattern of consequences at the large end. For the past ten years, the new transportation technology that has undergone the most assessment and debate has been the Supersonic Transport (SST).

Originally the controversy was over whether the United States government should continue support of the development of our own SST.

After much debate, the U.S. Congress voted to discontinue funding for the SST in 1971. About one billion dollars had been spent on the project. Meanwhile both the Russian government and a British-French development group continued work on their versions of the SST. A second controversy arose in 1975 when the British and French asked for permission to begin commercial test flights of their SST (Concorde) between U.S. cities and London and Paris. The then Secretary of Transportation, William Coleman, charged his staff to conduct an assessment of the impacts of test flights of the Concorde into and out of Dulles (near Washington, D.C.) and Kennedy (New York City) airports.

On February 4, 1976, Coleman announced his decision to allow a sixteen-month test period for the Concorde. At Dulles Airport the Concorde test flights began a few months after the decision, but at Kennedy strong local opposition resulted in a ban on Concorde landings.* From May, 1976, to October, 1977, a seventeen-month court battle carried the issue all the way. to the Supreme Court. On October 17, 1977, a Supreme Court decision cleared the way for daily commercial flights which began on November 22, 1977.

In order to understand the rationale behind Secretary Coleman's decision, the subsequent opposition, and the legal battles that are currently being fought, a summary of the assessment that was carried out by the Department of Transportation (DOT) is presented (15). The outline for doing a technology-initiated assessment that was described in section 4.2 will be used to summarize DOT's assessment of the SST.

Background of Assessment and Decision

The issue before Secretary Coleman was whether to permit the Concorde SST, manufactured jointly by the British and French, to operate in limited scheduled commercial service to and from the United States as follows: not more than four flights per day at John F. Kennedy International Airport (JFK), located on Long Island, New York, and not more than two flights per day at Dulles International Airport, located in Fairfax and Loudoun counties, Virginia.

On August 29, 1975, and September 21, 1975, respectively, British Airways and Air France applied to the Federal Aviation Administration

---

*On March 11, 1976, the Port Authority of New York and New Jersey announced that it was banning landings at Kennedy while it evaluated the operating experience of other airports used by the Concorde.

(FAA) for amendment of the respective operations specifications to allow Concorde flights to and from the United States. The Concorde is the first commercial transportation application of the new supersonic technology, a technology used for military aircraft for over twenty years. It is of European rather than American design and manufacture, unlike the vast majority of the aircraft in use worldwide today. The main advantage of the Concorde is that it halves overwater flight times. However, its impact on the environment is more severe than that of most existing commercial aircraft, the noise problem have the most impact on people who live near airports (see Table 4.4). Because a decision to admit the Concorde could constitute a major federal action significantly affecting the quality of the human environment within the meaning of the National Environmental Policy Act (NEPA) of 1969, the FAA prepared and released on March 3, 1975, a draft environmental impact statement analyzing the likely environmental consequences of permitting the aircraft to land in this country. On the basis of voluminous comments and continuing research, it then prepared the environmental impact statement that was released to the public on November 13, 1975.

Secretary Coleman also conducted a public hearing on January 5, 1976. Proponents and opponents of the Concorde presented arguments for over seven hours. Representatives of citizen groups, the manufacturers of the Concorde, the two foreign airlines involved, experts in

TABLE 4.4

NOISE LEVEL OF PLANES

(in effective perceived noise decibels)

Airplane	A-300 Airbus	727	DC-8	707	747	DC-10	Concorde
Gross weight (in tons)	173.6	97	175	158	389	250	200
Maximum capacity	250	140	260	189	400	254	100
No. of engines	2	3	4	4	4	3	4
Noise:							
—Takeoff	92	104	116	113	107	104	119
—Landing	101	108	117	118	106	108	117
Number in service	0	793	171	240	104	122	9

technology and the environment, and American, French, and British public officials addressed a series of relevant issues which highlighted the environmental, technological, and international factors that would have to be evaluated in reaching a decision.*

Impact Analysis

In attempting to decide on the desirability of using new technologies, all the relevant impacts have to be identified and studied. Then the beneficial impacts have to be weighed against the detrimental impacts to arrive at a decision. In the SST assessment the major impact areas were:
1. Speed and other performance benefits.
2. Impact on international relations.
3. Safety of Concorde flights.
4. Legal issues including international obligations and domestic law.
5. Environmental consequences:
    a. Noise pollution.
    b. Energy efficiency.
    c. Stratospheric impact.
    d. Air quality.

These impact areas were all studied by DOT's assessment team, and their 100-page report (15) provides detailed information about each of them.

4.5 TECHNOLOGICAL FORECASTING APPLIED TO TRANSPORT SYSTEMS

Assessment of technology requires accurate information about future potentialities and impacts. Technological forecasting is an integral part of the assessment process. When conducting a problem-initiated assessment, precise information about time frame and magnitude of alternative technologies is essential. The assessment of specific emerging or futuristic technologies requires a forecast of when the technology will be available and what the likely consequences of its usage are. The purpose of this section is to describe some of the more common methods of forecasting and to indicate their appropriate transportation applications.

Technological forecasting methods are customarily divided into three kinds, as indicated in Figure 4.4. *Exploratory* forecasting techniques start

---

*The newly proposed Science Court discussed in chapter 11 is intended to deal with technological disputes of this sort.

FIGURE 4.4

CATEGORIZATION OF METHODS OF TECHNOLOGICAL FORECASTING

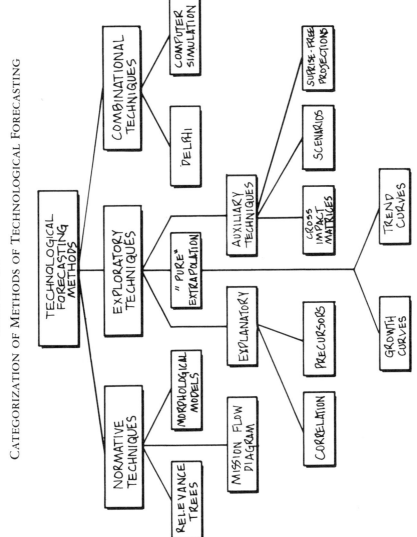

from the present situation and its preceding history and attempt to project future developments. *Normative* forecasts, on the contrary, start with some desired or postulated future situation and work backwards to derive feasible routes for the transition from the present to this desired future. *Combinational* techniques make use of information collected from the first two categories and integrate the data into a composite forecast (17). Here we shall study the normative forecasting methods.

Normative Forecasting Methods

The three most commonly used methods for preparing normative forecasts are *morphological models, relevance trees,* and *mission flow diagrams.* Morphological models are best suited for situations that can be broken down into pieces that are more or less independent and can be treated separately. Relevance trees are best suited for situations in which some kind of hierarchical dependence exists. Mission flow diagrams are best suited for situations involving some kind of flow, process, or sequence. Many situations, however, can be handled by more than one of these methods. The forecaster should select whichever method is most appropriate for the specific case at hand.

Constructing a *morphological model* of some situation involves identifying major elements of that situation, elements that themselves may involve several alternatives. A single representation of the situation then is synthesized by selecting one of the alternatives from each of the major elements. To illustrate this procedure, let us consider a mass transit system: the major elements of this system are vehicles, routes, and schedules. Note that riders here are specifically *not* considered as part of the system. The system exists to serve them. However, for other purposes, such as modeling an entire city, the riders might legitimately be considered as parts of the mass transit system.

Vehicles may be further described in terms of size, motive power, and type of control. Size means number of passengers, and we can take, as representative of a wide range of sizes, vehicles carrying five, thirty, or one hundred passengers. Alternatives for motive power might include diesel, steam, battery, turbine, and external electricity. The type of control might be manual, in which all vehicle functions (such as stopping at a station) are selected at the discretion of an operator but the function is then performed without operator control, or automatic, in which all functions are performed by the control system. Moreover, the control might be located either internally or externally. Thus, there are many possible vehicle types, each involving a selection of size, motive power, control type, and control location. Routes may involve some kind of guideway and may involve some degree of shared usage. Guideways may be fixed, like streetcar tracks; they may flexible, like the overhead wires of a trackless trolley, which permit the trolley to go anywhere there are overhead wires; or there may be no guideway at all, as with city buses. Regardless of the type of route, usage may be exclusive with the transit system (as is the case with streetcars, which have exclusive use of their tracks), shared (as with commuter trains which share their tracks with other types of trains), or general (as with buses which share the streets with all other types of traffic). Finally, the schedule may be historic, based on previous experience with numbers of riders at given hours on given days of the week (perhaps modified for seasonal conditions and special events); it may be responsive, as the transit management adjusts the number of vehicles on a route on the basis of whether demand is unusually heavy or light, or it may be real-time, if a vehicle is scheduled on a route only when it is demanded by one or more passengers. The major components of the system, and the alternatives for each, are listed in Table 4.5

TABLE 4.5
COMPONENTS OF MASS TRANSIT SYSTEMS

SCHEDULE

historic
responsive
real-time

ROUTES

Guideways	Usage
fixed	exclusive
flexible	shared
none	general

VEHICLES

Size	Power	Control Type	Control Location
5 passengers	diesel	manual	internal
30 passengers	steam	semiautomatic	external
100 passengers	battery	automatic	
	turbine		
	external		

Construction of a *relevance tree* involves subdividing some situation into finer and finer units, each included in or related to the units above it and containing the units below it.* To illustrate the use of relevance trees, another example will be examined. Moreover, to show the similarities and differences between relevance trees and morphological models, the example chosen will involve the same problem as the previous section, namely, mass transit.

Consider Figure 4.5, which is one possible relevance tree describing a mass transit system. The mass transit system is broken down into three

FIGURE 4.5

POSSIBLE RELEVANCE TREE FOR MASS TRANSIT

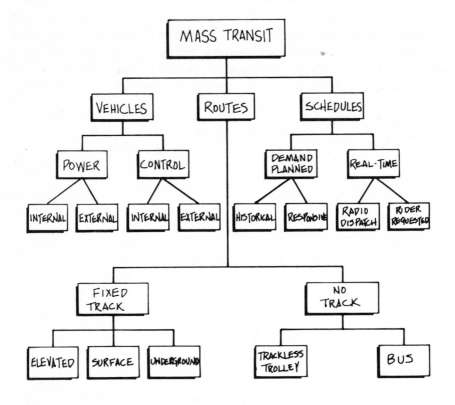

---

*The decision tree for categorizing energy policy alternatives depicted in Figure 4.1 is one example of a relevance tree.

major elements: vehicles, routes, and schedules. Vehicles, in turn, have two major elements: power and control. Power may be derived either internally or externally. Control may be located either internally or externally. If the tree were continued to more layers, each of these could be broken down into finer subdivisions. The second element, routes, may be categorized as either those involving tracks or those with no tracks. If tracks exist, they may be elevated, at surface levels, or underground. Each of these might be further subdivided into various types if the development of the tree were carried further. If the system does not involve tracks, it may require some other physical connection between the vehicles and the remainder of the system, as does the trackless trolley, or the vehicles may be physically independent of the rest of the system, as are buses. Finally, the schedules may be demand-planned prior to the dispatch of the vehicle from the yards or barns, or the scheduling may be real-time. In the demand-planned situation, scheduling may be based completely on historic data (perhaps modified for special events and conditions), or it may be responsive to observed demand, with vehicles being added or removed from a route according to the departure of an observed number of riders from the planned level. If scheduling is to be done on a real-time basis, riders may communicate with the dispatcher who then radios to a selected vehicle to make a particular stop. The dispatcher may select a vehicle that is already routed to go near the rider's pickup point and his destination, or he may re-route an idle vehicle. On the other hand, the scheduling may be done after the rider enters the vehicle, at which time he gives his destination to the driver, or otherwise communicates it to the control system.

Constructing a *mission flow diagram* involves determining all the routes or processes by which some function is performed and identifying each of the stops in the sequence along each path. This technique is especially useful in the analysis of transportation and communication networks.

4.6 EMERGING AND FUTURISTIC PASSENGER TRANSPORT SYSTEMS

This chapter concludes with a discussion of innovative attempts at better meeting current and future transportation needs of people. Transportation systems can be divided into two major classifications, intraurban and interurban. Intraurban systems are those systems providing transportation within a specified urban area. Interurban systems are transportation systems between urban centers. It immediately becomes apparent that commuter travel lies between the two categories and that some systems belong in both categories.

Further classification can be according to .demand, or user-initiated, systems as opposed to scheduled or operator-initiated systems. Demand systems generally fall within the intraurban classification, while scheduled-type systems may be either intraurban or interurban systems (7). Due to space limitations, only some of the innovative systems proposed for intraurban-demand use will be discussed in detail.

Intraurban-Demand Systems

The private automobile is the primary type of user-initiated transportation system that is used in urban areas. It is generally considered to be the most convenient means of personal transportation for the vast majority of adults. It is convenient both in terms of availability and flexibility. This convenience factor decreases, however, with the ever-increasing highway congestion and parking problems.

Automated Highways. The automated highway has been proposed as a means of increasing safety and at the same time increasing the capacity of superhighways. On such a highway, each vehicle would be fully controlled by the system both in speed and direction. This degree of control would permit a reduction in headway (the distance between vehicles) required for safety and an increase in speed. Fully automated highway systems, although already proposed and under investigation, may not be available for use until the 1990s. A United States Department of Transportation assessment entitled "the Northeast Corridor Transportation Project" recommends that research on automated highways be intensified (12). The assessment also says that proposed legislation for the Post-Interstate Highway Program should be prepared in such a way that highways will be planned and built to make it possible for them to accommodate future automation.

The most pressing issue related to the future of the private automobile is the increasing cost and decreasing supply of gasoline. Several alternative sources of fuel are being studied as possible substitutes for gasoline. Cars powered by methanol and hydrogen compounds in internal combustion engines are currently under investigation. The search for a suitable alternative power plant is also an important area of research.

One of the more promising options is the electric car.

Electric Cars. Many· companies have built or are experimenting with battery-operated electric vehicles. However, widespread use of such vehicles is limited by the size, weight, performance, and range characteristics of batteries. Hence current research and development have been focused on developing higher "energy density" batteries such as durable, higher-performance lead-acid batteries, while advancing the technology of zinc–nickel oxide batteries and lithium–iron sulfide batteries. The term "energy density" refers to the amount of energy that can be packed into a given weight of battery. Improved flywheels, to be discussed later in this section, are also being proposed as a substitute for batteries.

An official of Exxon Corporation, in discussing energy-density requirements, said: "The best lead-acid battery commonly available has an energy density of about 15 watthours per pound in electric vehicle applications. That imposes on a reasonable electric car of this day 30- to 40-mile range in typical urban use. For a 100-mile range, we require a battery with 30–50 watthours per pound energy density. Outside the laboratory, such batteries do not exist for practical use on a broad scale" (4). Thus, except in labs or in costly experiments (the moon buggy driven by American astronauts used a silver-zinc cell with energy density of 50 watthours per pound), lead-acid remains electric propulsion's only challenge to gasoline today. The general consensus is that a 30 to 50 watthours per pound battery will not be available until the early 1980s.

In terms of energy density, it's no contest. A 20-gallon tank of gasoline can provide about 2.4 million Btu's of energy. Lead-acid storage batteries weighing the same as the 20-gallon tank of gasoline can provide only about 7,700 Btu's of energy. Thus, gasoline has an energy density that is over 300 times that of lead-acid batteries. In total energy efficiency, a measure of the amount of energy it takes to move a car a mile, electricity leads gasoline—but not by much.

Current research efforts, besides searching for batteries with higher energy density, are also focusing on ways of improving the total efficiency of the electric car. For example, at General Motors, an effective regenerative braking system (which can return energy to the battery during deceleration) has now been developed, and an average of 10 to 15 percent improvement in electric vehicle range can be realized through its use.

From the late 1970s into the early 1980s, electric vehicles will still only be appropriate for limited applications. Current usage by the U.S. Postal Service and certain resort facilities gives an indication of increased future use in application areas with travel needs that are matched to the limited range and speed of electric vehicles. A fleet of 380 electric delivery vans is being used by the U.S. Postal Service and so far performance on routes averaging 11.5 miles has been satisfactory. It is estimated that there are some thirty thousand postal routes that fall within the parameters of speed, distance, and grade now available in present electric vehicles. At Sea Pines Plantation, Hilton Head Island, South Carolina, an electric vehicle called the Islander is rented to transient residents and tourists. It is capable of carrying four adults and 500 pounds of luggage at 30 mph for 50 miles.

The electric and hybrid vehicle (EHV) commercialization process is receiving a boost from a new Department of Energy project which has been mandated by the "EHV Research, Development and Demonstration Act of 1976" (P.L. 94-413). This act, which was amended on February 25, 1978, (P.L. 95-238), established an eight-year project to "demonstrate the economic and technological practicability" of using EHV's on the nation's roadways.

Dial-A-Ride: This type of system, which goes by many names, consists basically of a fleet of small buslike vehicles that are routed according to user demands. The size of the vehicles used depends on projected peak load estimates and can range from a passenger van to a regular bus. Usually a minibus which can carry about twenty passengers is used. Using vehicle-dispatcher communications en route, vehicles are rerouted as efficiently as possible to provide door-to-door service. In the mid-1970s, several demonstrations have been or are being conducted. Initial findings indicate that maintaining a sufficient ridership to offset the costs of purchasing equipment and operation is a major problem which needs to be solved.

Personal Rapid Transit (PRT). PRT is currently exciting transportation planners seeking a substitute for the automobile, which not only pollutes the atmosphere, but also is helping to exhaust the world's dwindling supplies of petroleum. This proposed transportation system consists of fixed-guideway systems in which automated vehicles no larger than small automobiles carry people and/or goods nonstop between any pair of stations in a network of slim guideways. The system can be designed to serve major activity centers such as business districts, shopping areas, and airports, or can span an entire urban area. Proponents cite PRT as an example of minimum-waste technology that can be designed to use a minimum of energy and material resources (1).

Intraurban-Scheduled Systems. A number of new systems and vehicles have been developed or are being proposed for intraurban-scheduled systems. By far the most ambitious urban mass transit project of the past twenty years (1957–1977) has been the BART (Bay Area Rapid Transit) system. Gordon D. Friedlander, in writing about the history of BART, said: "This continuing story can teach us much about the do's and don't's of applying technology on the grand scale" (5).*

In the early 1970's, the Department of Transportation awarded a $24 million grant to develop a new bus that could be used to serve the needs of urban travelers better. Working with three subcontractors, AM General, General Motors, and Rohr Industries, a new bus called "Transbus" was designed using the prototypes developed by the subcontractors. Unfortunately, the improved bus design results in a more expensive bus, and so far has not been able to compete with conventional urban buses.

Currently the major technological advances in urban mass transit have been the development of control systems for automated rail systems. For example, a joint venture of BG Checo Engineering Ltd. and Jeumont-Schneider of France is the design, manufacture, and installation of a $29 million ATO (Automatic Train Operation) system for the entire Montreal, Canada, Metro network. Cab signaling, with automatic train protection, will permit manual operation of trains when necessary; however, in the ATO model, the driver's control will consist of only two pushbuttons. One of these will control the train doors; the other must then be pressed for a predetermined time to put the train into a fully automatic program of starting, accelerating, coasting, and braking to a stop at the next station.

Flywheels in Transport Systems: New super-strength fibers and innovations in mechanical engineering have resulted in experimentation with flywheels that can store about ten times more energy than conventional stainless-steel flywheels. Flywheels are being used or proposed for use with subway trains, trolley buses, and automobiles.

The Metropolitan Transit Authority (MTA) of New York City is cur-

---

*BART was designed to relieve traffic congestion in the San Francisco Bay Area.

rently conducting an experiment with its subway cars which may indicate a new direction for mass transit. Two of the cars have each been equipped with two energy-storage machines. Each of these machines contains a set of four flywheels which are designed to save the energy normally lost as heat every time a train brakes to a stop at a station. Projections are that about a third, or $10 million, of the MTA's annual electric bill would be saved by a complete fleet of cars equipped with these energy-storage machines.

The savings may be achieved in the following way. As the train slows to a stop, the kinetic energy of the moving train is converted to the rotational energy in the flywheels. Having been revved to a maximum of 14,000 revolutions per minute (rpms), the flywheels spin freely in their vacuum chamber until the train begins to accelerate. Then the stored energy in the flywheel is transferred back to electric motors to get the train moving. This action lowers the rpms of the flywheels to a minimum 9800 rpms. The cycle is completed as the flywheels are recharged to their maximum when braking for the next stop. This process is called regenerative braking.

The steep hills of San Francisco, California, were the scene of a short-lived experiment with flywheel-run trolley buses. The Lockheed Corporation outfitted buses with flywheels that were revved up at each stop while passengers were loading and exiting. The electricity was transferred by means of overhead contact; a complete "charge" took less than a minute. The flywheels allowed the pollution-free buses to travel five to six miles between charges.

This is a tremendous advantage over other trolley buses which must remain attached to unsightly overhead wires for the whole length of their route. When accidents or fires blocked their path, this type of trolley was held up, and so was all the service along the route from that point on. But flywheel-equipped trolleys are able to be rerouted around road blocks with little or no disruption of service.

The San Francisco project was curtailed in 1976. Worry about what would happen to passengers if the vehicle were involved in a collision was the reason. If normal control of the vehicle were interrupted, the rotational motion of the flywheel would take over, causing the bus to spin.

A significant percentage of city air pollution is caused by the automobile. However, no one who understands Americans' dependence on the auto seriously suggests banning it from the road in the foreseeable future. Further modifications of the car are needed to cut down on air pollution. Changes are also needed to decrease the gasoline consumption of automobiles.

Some engineers believe that flywheels may be the ultimate answer. They see a flywheel car with a cruising range of 40 miles being available by 1990. Longer trips would be possible by means of a hybrid system—a regular gasoline engine would rev the flywheel to its peak rpms. Like the subway and trolley bus systems, some recharging of the flywheel could be done by "plugging" the car into a 220-volt electric line. This plan is currently being worked on at the Stanford Research Institute.

Concluding Comments

In considering ways of overcoming transportation problems, besides looking for improved transportation technologies for meeting travel needs, we also need to look at the possibilities for substituting telecommunications for transportation. In an NSF-supported assessment called "Development of Policy on the Telecommunications-Transportation Tradeoff," Jack M. Nilles and his colleagues said:

> Given the capability of modern telecommunications and computer technologies to efficiently produce, transmit, and store information, it appears probable that many information industry workers could "telecommute." That is, they could perform their work, using communications and computer technologies, at locations much closer to their homes than is the case now. (13)

The telecommunications-transportation trade-off is another example of a complex sociotechnological issue that can only be resolved by careful analysis. The business and technological environments that make change possible need to be explored. Finally, the social, environmental, energy, and other impacts that affect the problem have to be studied.

As new transportation options are proposed, designed, and field-tested, an ongoing assessment must be carried out to provide the public, professional transportation planners, and public officials with the information that is needed to make wise choices. We need to match our transportation systems better with our transportation needs. We need to match our transportation systems better to human users. We also must continue our search for minimum-waste alternatives that have a minimum impact on our natural and people-made environments.

4.7 EXERCISES

1. Find out whether your regional area is also experiencing the trend of decreasing usage of mass transit. If yes, what is being done or being proposed to reverse this trend? If no, what has been done to counter the national trend of decreasing ridership of public transit systems?

2. What is unique about technology assessment (TA) that separates it from other ways of studying technology?

3. Comment on the reasons and trends discussed in the section entitled "Is TA Needed?" Explain why you agree or disagree with the rationale given for the need for TA studies. Cite another trend or reason, not discussed in the chapter, which makes TAs necessary.

4. Seven possible consequences following a TA are listed in section 4.2. Cite an example for each of the listed follow-up actions to a TA.

5. What is the basic difference between problem-initiated and technology-initiated assessments? Cite an example, other than those discussed in the chapter, for each type of TA.

6. In Figure 4.1, energy policy alternatives for conserving energy used by passenger transportation systems are outlined. Specifically, options for decreasing energy intensiveness are discussed.
   a. Discuss the techniques and impacts associated with the alternatives for saving energy by decreasing travel, namely: increasing the cost of gasoline, rationing gasoline, and changing housing patterns to decrease the need for travel.
   ·b. Develop the branches of the decision tree associated with freight or recreational transportation systems.

7. In Figure 4.2, both the historical variation and differences in energy intensiveness (EI) of various passenger modes of travel are depicted. Social, economic, consumer, and other factors affect the choices of passenger transport that result in these usage patterns.
   a. Discuss some of the reasons for the general increase in energy intensiveness (EI) from 1950 to 1970. Why was there such a large increase in EI of airplane travel while railroad EI decreased substantially?

b. In 1970, urban use of the automobile was about twice as energy-intensive as intercity use. What are some of the factors responsible for this difference?

8. In Figure 4.3, ferry service choices and consequences are outlined for organizing a study to explore ways of increasing and improving ferry services across Long Island Sound. Select a mass transportation need or problem in your regional area and develop a diagram similar to Figure 4.3.

9. Conduct a library search to update the legal battles related to former Secretary of Transportation Coleman's decision to allow trial service of the Concorde SST at Kennedy International Airport.

10. If you live or go to school near an international airport, conduct a survey to determine the opinions of community groups concerning potential Concorde SST service in your locale. Groups representing the Miami, Florida, and the Houston, Texas, areas have already expressed an interest in having SST service at their respective airports.

11. Select a futuristic transportation technology of your choice and conduct a library search for a technological forecast of the technology chosen.

12. The General Electric Company is currently developing a polar woven flywheel made from tailored composite materials. This GE-RESD flywheel is designed to achieve high-energy densities of the order of 40 to 60 watthours per pound as compared to 10 watthours per pound for steel flywheels.
    a. Compare the energy densities achievable by the GE-RESD flywheel to the batteries that are under development for use in electric cars.
    b. What are the advantages and disadvantages of using a flywheel instead of a bank of batteries in future electric cars?

13. Write an essay agreeing or disagreeing with the thesis that in order to increase the ridership on mass transit, new public transit systems must have as many of the characteristics of a private automobile as is possible. This premise implies that the additional expense needed for making mass transit more convenient, private, comfortable, etc., is worthwhile and necessary.

4.8 BIBLIOGRAPHY

1. Anderson, J. E. "PRT: Urban Transportation of the Future?" *The Futurist*. vol. 7, no. 1, February 1973.
2. Caldwell, W. A., ed. *How to Save Urban America*. A Regional Plan Association Book. New York: Signet, 1973.
3. *Crossing the Sound*. A study by the Tri-State Regional Planning Commission, One World Trade Center, New York. December 1975.
4. Dedera, D. "Electrifying the Family Car." Exxon USA, Public Affairs Department of Exxon Corporation. 1976.
5. Friedlander, G. D. "The BART Chronicle." *IEEE Spectrum*. September 1972.
6. *The Futurist*. vol. 5, no. 6, December 1971; vol. 6, no. 1, February 1972. Special issues of the journal of the World Future Society, on the topic of technology assessment.
7. Glorioso, R. M., and F. S. Hill. *Introduction to Engineering*. Englewood Cliffs, N.J.: Prentice-Hall, 1975.
8. Hirst, E. "Transportation Energy Use and Conservation Potential." *Science and Public Affairs*. November 1973.
9. Kieffer, J. A. "The Automobile's Success: A Lesson for Its Critics." *The Futurist*. vol. 7, no. 1, February 1973.
10. Kirchmayer, L., et al., eds. *A Technology Assessment Primer*. Institute of Electrical and Electronics Engineers, 1976.
11. *Mobility: From Here to There*. Number 5 of "The Markets of Change" series, Kaiser News, 1971.
12. Myron, M., et al. "Recommendations for Northeast Corridor Transportation." Strategic Planning Division, Office of Systems Analysis and Information, U.S. Department of Transportation, 400 Seventh St. S.W., Washington, D.C. 20590.
13. Nilles, J. M., et al. *The Telecommunications-Transportation Tradeoff*. New York: Wiley, 1976.
14. Rose, David J. "Energy Policy in the U.S." *Scientific American*. January 1974.
15. "The Secretary's Decision on Concorde Supersonic Transport." Department of Transportation, Washington, D.C., February 4, 1976.
16. *A Study of Technology Assessment*. Report of the Committee on Public Engineering Policy, National Academy of Engineering, 1969.
17. World Future Society. *An Introduction to the Study of the Future*. Washington, D.C., 1977.

Chapter Five

COMMUNICATIONS

5.1 INTRODUCTION

The methods of long-distance communication from prehistory until the invention of the electric telegraph changed very little. People communicated by visual means such as the optical semaphore, by the sound of drums, gongs, and trumpets, and most frequently by messengers.

Many pre-technological societies developed extensive communication networks extending over thousands of miles. But the existing techniques limited the speed with which messages could be transmitted. During the height of the Roman Empire, for example, a messenger traveling over the great Roman roads could deliver a message from Rome to London in thirteen days. In the 1830s, more than fifteen hundred years later, the delivery of a message from Rome to London still took the same length of time.

The development of the electric telegraph in the 1840s and 1850s revolutionized communication. With the installation of a telegraph line, almost instantaneous communication was established over thousands of miles. The subsequent development of the telephone, radio, and television resulted in a global communication network that was unimagined even thirty years ago. Today there is hardly an aspect of modern society that has not been influenced by our communication technology.

In this chapter we will describe the basic principles of modern communication. We will explain the operation of the telegraph, telephone, radio, and television, and we will describe some future trends in the technology. In the final section we will briefly examine the impact of communication on various aspects of society.

Modern communication is based on the application of electromagnetic phenomena. The study of communication technology must therefore begin with an explanation of electricity and magnetism and their interaction.

5.2 ELECTRICITY AND MAGNETISM

The first reports of electric phenomena come from ancient Greece. It was known then that when amber is rubbed against a cat's fur, the amber attracts small objects before touching them. This property of amber disappears a few minutes after the rubbing. The phenomenon was thought to be unique to amber and was named electricity after the Greek word for amber, "elektron." A more substantial knowledge of electricity was not obtained until about the year 1600, when William Gilbert showed that electricity is not a unique property of amber but that it is manifest in many other substances.

By the middle of the 1700s, through the work of many researchers, a large amount of phenomenological knowledge was obtained about electricity. But not until the early part of this century were these electrical phenomena correctly explained in terms of the properties of matter.

All electric phenomena result from the interaction of electric charges, which are a basic property of matter. There are two types of electric charge: positive and negative. Charges exert forces on each other, such that like charges repel each other and unlike charges attract each other. *Atoms* which make up matter contain small negatively charged *electrons* and relatively heavier positively charged *protons*. The proton is about a thousand times heavier than the electron, but the magnitude of the charge on the two is the same. There are as many positively charged protons in an atom as negatively charged electrons. The atom as a whole is therefore electrically neutral. The identity of an atom is determined by the number of protons it has. For example, hydrogen has one proton, carbon has six protons, silver has forty-seven protons. Through a series of ingenious experiments it was shown that most of the atomic mass is concentrated in a *nucleus* made up of the protons. The electrons orbit around the nucleus much as the planets orbit around the sun. They are maintained in orbit by the electrostatic attraction of the nucleus.

This simple model of the atom is not complete; still it does provide a qualitative explanation for many of the electric properties of matter. Objects can be electrically charged by rubbing because when two objects are brought into close contact, the surface electrons from one object enter the surface of the other. When the two objects are rubbed against each other, electrons are removed from one object and deposited on the other. In this way when amber is rubbed by fur, electrons are removed from the fur and are deposited on the amber. Amber, with excess electrons, therefore becomes negatively charged and since those electrons have been removed from the fur, it is left with an excess positive charge.

Early experimenters showed that some materials conduct electricity while others do not. This can be explained at least qualitatively by the type of interatomic bonding that exists in the material. In a conducting material, such as copper or silver, some of the electrons are held rather loosely to the nuclei. These electrons do not specifically belong to any single atom but travel freely throughout the material. In insulating materials, such as rubber or sulfur, the electrons are tightly bound and localized. They are not free to move within the material.

Prior to the 1600s the understanding of *magnetism* was on about the same level as the understanding of electricity. In fact, not much distinction was made between the two. The Greeks discovered that a certain kind of iron oxide (called magnetite or lodestone) attracted or repelled other pieces of the same material or small bits of iron. When a piece of magnetic material was freely suspended, one end of it pointed north. It is not known exactly when this effect was discovered, but it was used as a navigational aid by the twelfth and thirteenth centuries.

Many of the experimenters who did the early work on electricity also studied magnetism. They found many similarities between electric and magnetic phenomena. Just as in electricity, in magnetism there also are two kinds of entities, the *north* and *south poles,* and both attraction and repulsion can be observed between magnetic materials. It is again observed that like poles repel and unlike poles attract each other. There are, however, some important differences between electricity and magnetism. It was William Gilbert around the year 1600 who made the clear distinction between magnetism and electricity. Although positive and negative electric charges can exist separately, the north and south magnetic poles cannot be separated. If a magnetic bar is broken, the opposite north and south poles appear at the point of fracture.

An electric charge exerts a force on another electric charge; and a magnet exerts a force on another magnet. These two forces have an important common characteristic: the exertion of the force does not require physical contact between the interacting bodies. The forces act at a distance. A very useful way of visualizing these forces which act at a distance is the concept of lines of force or field lines which extend beyond the electric charge or the magnet. Thus an electric charge is surrounded by an *electric field* and a magnet is surrounded by a *magnetic field*. The electric field exerts a force on an electrically charged object while a magnetic field exerts a force on a magnet.

Through Gilbert's work it was known that electric and magnetic phenomena are different. Still, most experimenters including Gilbert suspected that there was a connection between them. Attempts were made to establish the relationship between the two effects. Charged objects were brought near magnets, but no mutual force was observed between them. The connection between electricity and magnetism was not established until 1820. In retrospect it is not surprising that this took so long. Stationary charges do not affect magnets. In order to interact with a magnet a charge has to be in motion. Charges in motion are called *electric current*. In the early electricity experiments, the charges were generated by friction and they were mostly stationary. A sudden motion of charges could be produced by allowing charges to "jump" across a small gap (producing a spark) but this current was small and of short duration. A steady source of electric current was not available until the 1790s when Alessandro Volta invented the electrochemical battery.

The battery is a device in which a chemical reaction involving two terminals made of dissimilar metals produces a continual removal of electrons from atoms. Due to this chemical reaction a charge difference is produced between the terminals. A continuous flow of charges (current) is obtained in a conducting material that connects the two terminals.

With the invention of the battery the stage was set for the discovery of the long-sought connection between electricity and magnetism. The first half of the connection was found in 1820 by a Danish scientist, Hans Christian Oersted.

Oersted's great discovery was reportedly accidental. As the story is usually told, he was lecturing to his students on electricity and magnetism. As part of the demonstration he connected the two terminals of the voltaic battery to produce a current flow. By chance there was a magnetic compass needle near the wire, and he observed that the needle rotated to a position at right angles to the current-carrying wire. When he reversed the connections to the battery (thus reversing the motion of the charges), the needle reversed its direction but remained perpendicular to the wire.

Oersted realized the importance of this effect: he had found the connection between electricity and magnetism. The current-carrying conductor produces a magnetic field and therefore behaves as a magnet. Such electromagnets produced by current-carrying conductors have some very useful properties not present in permanent magnets. The polarity of the electromagnet can be reversed by changing the direction of current flow and the magnet can be turned off simply by switching off the current.

Since a current produces a magnetic field, it is logical to ask if a magnetic field can produce a current. This question was answered by Michael Faraday in 1831. He showed that a moving magnet generates (induces) a current in a conductor. The current flows as long as there is relative motion between the magnet and the conductor. When the motion stops, so does the current.

5.3 TELEGRAPH

Following the discoveries of the relationships between electricity and magnetism, progress in communications was very rapid. One of the first electromagnetic telegraphs that operated over a significant distance was built in 1834 by Gauss and Weber, two scientists at the Gottingen Observatory.

They strung insulated copper wire over the rooftops of Göttingen, connecting the astronomical observatory to the magnetic observatory over a mile away. The signaling current was obtained from voltaic batteries and, on occasion, from a small electromagnetic generator. The receiver consisted of a magnet, free to swing inside the coil carrying the signal current. The direction in which the magnet swung depended on the direction of the current in the coil. They developed an alphabet code in which the letter L. for example, was signaled by two swings of the needle to the left followed by a swing to the right.

There were many others who made significant contributions to the technique of electromagnetic telegraphy and built working telegraphs; but during its early stages the telegraph, like most other inventions, was a cumbersome gadget with many unsolved technical problems. In order to obtain financing and adoption of such a system, potential investors and backers had to be convinced of the value of the invention. The commercial adoption of the electromagnetic telegraph was pioneered by Cooke and Wheatstone in England and Morse in the United States. The more versatile Morse system eventually gained worldwide acceptance.

Although entrepreneurs like Cooke and Morse are indispensable to the development of a technology, it must be remembered that the devices they champion are almost always based on a large body of scientific and technological research to which many people contributed.

The basic principle of telegraphy is simple. A coded current is sent through a pair of wires from the sender to the receiver, where the message is decoded. (Two wires are necessary because a return path for the current must be provided.) This is illustrated by the Morse telegraph shown in Figure 5.1. At the sending end, key A is depressed, which allows a current to flow through coil B at the receiving end. The magnetic field produced by this current attracts the pivoted stylus holder and causes the stylus to write on the moving paper strip. By interrupting the connection with key A, the "dot" and "dash" signals are transmitted and recorded. Some of the Morse code symbols are shown in the figure.

FIGURE 5.1

MORSE TELEGRAPH

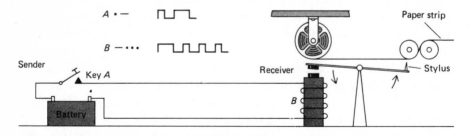

When Key A is depressed, the electromagnet B at the receiver is energized and attracts the pivoted stylus arm. The stylus then draws a line.

As the use of the telegraph increased, it quickly became evident that it was not practical to have separate pairs of wires connecting each user of the system. By the 1860s central telegraph offices were built in major cities. Each local telegraph station was connected to the central office in its region by one pair of wires. The local office sent its message to the central office, which relayed it to its destination. Depending on the destination, the message may have passed through a number of centrals before reaching the final receiver. In this simple and obvious way a telegraph office could communicate with all other offices in the world.

Toward the end of the last century, teletypewriters were developed that increased the speed of transmission and decreased the number of human operators necessary to handle the system. Using a teletypewriter, the operator depresses a typewriterlike key on the keyboard. This activates a set of electromagnets called relays, which automatically switch on electric currents and generate the electric pulse sequence corresponding to the letter on the key. At the receiving end these pulses energize a set of relays which then cause the transmitted letter to be typed on paper.

The early teletypewriters, of course, required an operator to type the letters. The system was not as efficient as it could have been because the machine was able to transmit messages much faster than the operator could type them. A great increase in the speed of transmission was brought about by the introduction of punched tapes. The operator types the message on a machine that perforates a paper tape with holes. A different combination of holes across the tape represents each letter of the alphabet. The perforated tape is then fed into a transmitter machine in which the holes in the tape are probed by metal fingers. The fingers close electric circuits through the perforated part of the tape, producing the pulsed code signal for transmission. At the receiving end the message is either typed by a teletype machine or stored on tape for later display. The speedup of transmission occurs because a tape transmitter can transmit messages at a rate more than twenty times faster than the transmission rate of a manually operated teletypewriter. One tape transmitter can therefore process tapes from many operators.

The speed and capability of the system were further improved by the introduction of automatic switching between transmitting machines. This allowed messages to be quickly routed from one station to another. Recently computers have been introduced into the telegraph system. The message, instead of being recorded on perforated paper tape, is stored in the computer. The computer automatically groups the messages and then feeds them to the transmitter. In many installations the telegraph system has been made completely automatic.

When the telephone was first introduced into public service, it appeared for a while that telegraphy might be completely replaced by the newer device. This did not happen. The technological improvements in telegraphy, some of which we have just described, gave the telegraph two important advantages over the telephone. Messages can now be transmitted at a much higher rate by telegraph than by telephone and a permanent record of the message is produced automatically. This is especially important in cases where the message concerns a number of people. Today the telegraph is widely used in news, diplomatic, and business message transmission. In a typical application, for example, each branch office of the user is equipped with a teletypewriter through which messages can be relayed and received to and from the central office. In addition, through an automatic central office the local office can contact any other teletype installation in the world. The teletype facilities of the U.S. State Department often receive as many as seven thousand diplomatic messages a day. The printed messages are automatically duplicated and distributed to an average of one hundred people for evaluation. Clearly such efficiency would be hard to obtain with a telephone.

5.4 TELEPHONE

Sound

With the telegraph completed, inventors' interests next turned to the problem of transmitting sound. Sound is a mechanical phenomenon produced by vibrating objects. When an object such as a tuning fork or the human vocal chords is set into vibrational motion, it transfers this motion to the surrounding air molecules, causing alternate compression and rarefaction of the surrounding medium. These pressure variations are transferred to adjacent regions, resulting in the propagation of sound away from the source.

Communication directly by sound is inefficient. The propagating medium has to be set into vibrational motion and because the propagating materials are not perfectly elastic, the motion is quickly dissipated by friction. With the discoveries of Faraday in 1831 it became possible to translate the information content of a sound wave into an electric current signal, which could then be transmitted by conducting wires to its destination. There it could be converted back into sound. Because the loss of signal strength is smaller, an electric signal can be conducted over much larger distances than sound.

Principles of Telephony

The conversion of sound into an electric signal is based on the discovery of Faraday that a moving magnet induces in a coil a current that is proportional to the motion of the magnet. This is the operating principle of a microphone (Figure 5.2). A membrane is attached to a magnet which can move freely inside a coil of wire. Pressure variations of the sound cause the magnet to move back and forth inside the coil. Because the motion of the magnet is proportional to the pressure variations caused by sound, the current induced in the wire is also proportional to the sound. An intense sound produces a large displacement of the magnet and therefore a large current. Similarly, a low sound produces a small current. In this way the fluctuations of pressure in the sound wave are transformed into corresponding fluctuations of current. This current is then conducted to the receiving station, where it is converted back to sound by a speaker.

The conversion of an electric signal into sound is based on Oersted's discovery that a current produces a magnetic field. The basic construction of a speaker is the same as the construction of a microphone.(Figure 5.2). The signal current flows through the coil and produces a magnetic field that causes the diaphragm to move in the field of the permanent magnet. The motion of the diaphragm follows the current variations and sets up pressure waves that are audible. Since the speaker diaphragm duplicates the motion of the microphone magnet, the sound it produces is identical to the driving sound.

The commercial application of the telephone was pioneered by Alexander Graham Bell, who built his first telephone in 1876. In 1877 Bell established the Bell Telephone Company and a year later the first telephone exchange was opened in New Haven. The telephone developed rapidly in many countries. In Britain the first commercial telephone lines were established in 1878 using mostly equipment imported from the United States. From the very beginning in Britain the telephone development was under government control by the general post office. Through investments of the American Telephone and Telegraph Company, which was set up by the Bell interests in 1885, telephone communications were built in France, Belgium, Spain, and Germany.

There was still a limitation to telephone communication. It was restricted to relatively short interurban distances. The electric currents generated by the microphone became too weak to drive the telephone speakers at the receiving end after traveling a few hundred miles. This problem was solved by the development of electronic amplification. In 1915, using newly developed amplifiers, a telephone link was established between New York and California.

The first transatlantic telegraph cable was laid in the 1860s. It took nearly another hundred years for a transatlantic telephone cable to be

FIGURE 5.2

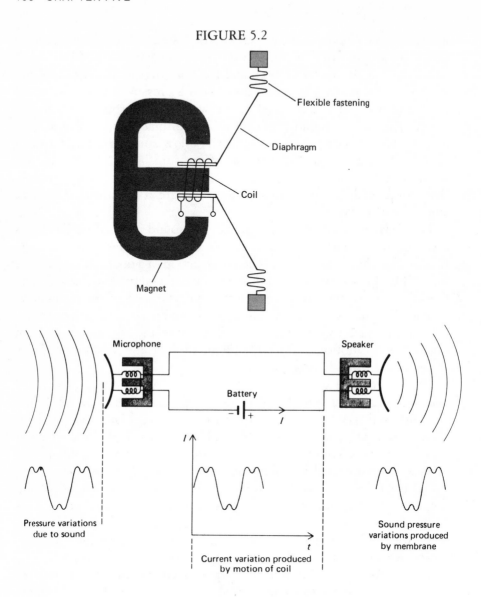

A. MICROPHONE. Pressure variations due to sound produce motion of the diaphragm and the attached coil. The coil moves in a magnetic field and therefore a current proportional to the sound is induced in the wires. In principle the construction of a speaker is identical to that of a microphone. Here a current applied to the coil causes the coil and the attached diaphragm to move. The diaphragm produces motion in the surrounding air.

B. THE OPERATION OF A TELEPHONE.

laid. The reason for this was the lack of suitable amplifiers. As we have mentioned, a telephone signal propagating along a wire has to be amplified at hundred-mile intervals if it is to remain strong enough to operate a receiver. Although amplifiers were already available in the early 1900s, they required much more development before they were adequate for underwater operation. Since a failure of an amplifier in an underwater cable requires the pulling up of at least a section of the cable, repairs are very expensive. The underwater amplifier must therefore be very reliable. Underwater amplifiers in current use are expected to operate at least twenty years without servicing. In addition, the amplifier has to be watertight and able to withstand the great water pressure under the sea. The first telephone cable connection between North America and Europe was established in 1953. Today there are hundreds of underwater telephone cables reliably interconnecting all parts of the world. Since the late 1960s communication satellites have further expanded the long-distance telephone networks.

A central exchange is required for telephones for the reason we discussed in connection with the telegraph. The first exchanges were manually operated. The person making a call picked up the receiver which operated a pair of contacts causing a light to flash in the exchange. The operator then spoke to the sender and activated the bell of the telephone at the receiving end. When that telephone was answered, the operator plugged in the sender's wires to those of the receiver and the call was completed. This system was adequate for as long as the number of subscribers was relatively small. But as the number of telephone users grew and electronic amplification increased the range of telephone contact, the manual switchboards were not able to cope with the load. By about 1912 automatic switching techniques were developed and introduced into some of the busy central exchanges. This innovation increased enormously the capacity of the telephone system. The dialing of the numbers opened and closed electric contacts which generated pulses that were transmitted to the central switching station. Here the signals activated the appropriate relays to complete the connection to the telephone with the dialed number. Although in principle the system is simple, the switching is, in fact, very complex because of the large number of subscribers tied into the network. The system must provide for alternate routing of the call when the most direct connection is used by another subscriber. It must also first signal the subscriber that the line is open for dialing (the dial tone) and then indicate whether the connection is completed or not (the ring or busy signal). The system has been further complicated by the introduction of direct long-distance dialing and other services provided by the telephone company.

The electromechanical switching systems are now being replaced by an entirely electronic system. The relays are being replaced by solid-state switches which operate without mechanical contact. The new system is faster and less noisy than the older one. Since response of the solid-state switches is faster, the conventional dial is replaced by push-buttons that can generate the number code more rapidly than the dial. Computers which have been recently introduced into the system have reduced the cost of billing and have further increased the efficiency of the telephone system.

A new device in telephony is the *videophone*. As the name implies, it is a combination of telephone and television. At both the sending and receiving terminals of the system a television camera and receiver are coupled to a conventional telephone. Those conducting the conversation are televised and the visual image is transmitted and displayed simultaneously with the sound. The basic technology for this system is already developed. Most of the effort is now concentrated on making the system more economical and versatile.

5.5 WIRELESS COMMUNICATION

Basic Principles

The theoretical foundation of the radio was put forth in 1864 by James Clerk Maxwell, the great theoretical physicist of the nineteenth century. Maxwell's work in electromagnetism was purely theoretical. He examined all the electromagnetic phenomena discovered by Oersted, Faraday, and others and formulated mathematical equations to explain them all in a unified way. His equations summarized all the known electromagnetic laws and interactions and in addition predicted some new important effects. Maxwell's equations can be summarized as follows: *a changing magnetic field creates an electric field and a changing electric field creates a magnetic field.*

Based on the equations for the electric and magnetic fields, Maxwell made a most important prediction. He showed that it should be possible to produce an electromagnetic field that propagates away from the source. Such a propagating field is called an electromagnetic wave. The process is illustrated in Figure 5.3. Let us consider a charge at the end of a rod. When the charge is stationary, an electric field emanates from the charge; there is, however, no magnetic field around the charge. But when the charge is put in motion, a current is produced by the moving charge and therefore the charge is surrounded by a magnetic field. If the charge has an oscillatory motion, then within each cycle the current and the associated magnetic field reverse direction. The changing magnetic field produces an electric field that is perpendicular to the magnetic field. Since the source of this electric field is continuously changing, the electric field is also changing, which in turn produces another magnetic field. In this way a propagating electromagnetic wave is produced in which the changing electric field produces the changing magnetic field, which in turn produces another changing electric field, and so on. The electromagnetic wave is in this way self-reproducing and detached from the source. The frequency of the field, or the number of times per second that the electric and magnetic fields undergo a complete change, is determined by the oscillating frequency of the source change. Maxwell showed that in empty space the electromagnetic waves propagate at a constant speed independent of their frequency. On the other hand, in a material medium the speed of propagation depends on the frequency and is generally less than the speed in empty space.

FIGURE 5.3

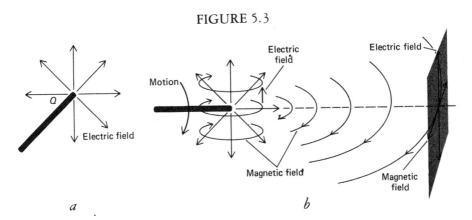

a   *b*

(*a*) A stationary electric charge is surrounded by an electric field but no magnetic field. (*b*) If the charge is set into motion, a current results which produces a magnetic field. The back-and-forth motion of the charge results in a changing current and changing magnetic field. The changing magnetic field produces a changing electric field, which in turn produces a changing magnetic field, and so on. A small part of the electric and magnetic field detaches from the source and propagates into space.

At the time that Maxwell developed these ideas many of the properties of light were already well known. It was known, for example, that light travels more slowly in materials such as glass than it does in empty space. The various interference properties of light such as the light-dark alternating regions produced when a beam of light passes through a narrow slit were also known. Maxwell pointed out that his postulated electromagnetic waves exhibit all the properties of light and he suggested that light is also an electromagnetic wave. The characteristic speed at which electromagnetic waves propagate is therefore just the speed of light, which was by then known to be about 186,000 miles per second.

When Maxwell first published his equations governing electromagnetic phenomena, there were very few people who understood them. His work was elegant and concise but way ahead of his time. The theory was fully explained and verified by Heinrich Hertz (1857–1894) in 1887, more than twenty years after its publication. At that time the key point in Maxwell's equations was the prediction that light and electromagnetic waves of the type generated by oscillating currents were the same phenomenon. It is this that Hertz set out to prove. Later in his career in a popular lecture Hertz told his audience, "I am here to support the assertion that light of every kind is itself an electrical phenomenon. The light of the sun, the light of a candle, the light of a glowworm."

Hertz's equipment is shown in Figure 5.4. When switch A is opened a large electric field causes charges to jump across the gap and to oscillate back and forth across it. As the current flows across the gap it heats the air in the gap causing it to glow. In this way the path of the current is

FIGURE 5.4

HERTZ'S EXPERIMENT

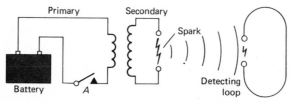

The oscillating current in the spark produces electromagnetic radiation which in turn produces a small spark in the detecting loop.

visible as a spark. The pulse of oscillating current generates a pulse of electromagnetic field, a portion of which propagates away from the spark gap.

The detection of the electromagnetic wave is simple in principle. If a conductor is placed in the path of the propagating field, the electrons in the receiving conductor are forced into motion, producing a current that can be detected. For detection Hertz actually used a loop of wire with a small gap in it. He detected the current caused by the field by observing a spark in the gap.

With this simple apparatus Hertz demonstrated that electromagnetic radiation produced by the motion of charges behaves as light. By putting a conducting surface in the path of the electromagnetic wave he showed that the wave is reflected by the surface as light is reflected by a mirror. He also demonstrated that electromagnetic radiation can be focused and that it propagates at the speed of light.

What then is the difference between light and radio waves? We certainly know that there are differences in their behavior. For example, light is blocked by objects through which radio waves penetrate without any difficulty. We can see objects that are illuminated by light, but our eyes are not sensitive to radio waves. We might also include in this discussion the difference in properties of x rays, cosmic rays, infrared rays, and microwaves—they all are electromagnetic radiation. The different behavior of these radiations is due entirely to the difference in their frequencies.

The whole spectrum of electromagnetic radiation is shown in Table 5.1. Electromagnetic waves, and all other waves, are characterized by three parameters, *frequency, wavelength,* and *speed of propagation.* The frequency is simply the number of times the field changes direction through a full cycle in one second. We will use the symbol f for frequency. The wavelength is a distance the wave travels during one cycle. Traditionally the Greek letter λ (lambda) is used to designate wavelength and c the speed of propagation. The parameters c, λ, and f are related by the equation $\lambda f = c$.

We can see from Table 5.1 the enormous range of electromagnetic wavelengths found in nature. The wavelengths extend from the unimaginably small 10^{-22} centimeters to many kilometers. We know from Maxwell that all of these wavelengths are produced by the same phenomena, namely accelerating or oscillating electric charges. The wavelength of electromagnetic radiation depends on the distance over which the source charge oscillates (moves back and forth). It is always longer

TABLE 5.1

THE ELECTROMAGNETIC SPECTRUM

f Frequency (sec^{-1})	λ Wavelength (cm)		
10^{32}	3×10^{-22}		
10^{30}	3×10^{-20}	Cosmic	
		ray	
10^{28}	3×10^{-18}	photons	
10^{26}	3×10^{-16}		
			$\leftarrow$ Limit of accelerator energy
10^{24}	3×10^{-14}		$\leftarrow$ Wavelength = size of elementary particle
10^{22}	3×10^{-12}		$\leftarrow$ Limit of nuclear gamma rays
10^{20}	3×10^{-10}	Gamma rays	$\leftarrow$ Limit of atomic rays
10^{18}	3×10^{-8}	x rays	
10^{16}	3×10^{-6}	Ultraviolet	Visible light
10^{14}	3×10^{-4}	Infrared	
10^{12}	3×10^{-2}	Microwaves	
10^{10}	3	Radar	
		UHF	
10^{8}	300	VHF, FM	
		Shortwave	
10^{6}	3×10^{4}	AM radio	
10^{4}	3×10^{6}	Longwave	
		radio	
10^{2}	3×10^{8}		$\leftarrow$ Wavelength = radius of earth

than the distance of oscillation. Thus charges oscillating in an antenna with dimensions on the order of centimeters produce radio waves, but light waves can be produced only if the charges are moved through distances shorter than the wavelength of light, which is about 5×10^{-5} centimeters. It is not possible to manufacture an antenna this small; however, the motion of electrons within the atoms is on this order. Electromagnetic radiation in the x-ray and light wavelength regions is emitted by electrons within atoms. The still shorter wavelength gamma rays are produced by the motion of charges within the nucleus.

Beginning of Wireless Communication

Although Hertz himself did not think that his findings were of any practical use and argued against the possibility of using electromagnetic waves for communication, there were many people who saw immediately the potential in Hertz's demonstration. Electromagnetic waves are, in fact, an excellent means of communication. Wire connections are not needed between sender and receiver and the speed of communication is nearly instantaneous.

The principle of using electromagnetic waves for communications is actually contained in Hertz's experiment. It is summarized in Figure 5.5. At the transmitting·end, charges are set into oscillatory motion in a conductor called a transmitter antenna. The oscillating charges produce an electromagnetic field, a portion of which radiates into space. At the receiving end, the electric part of the electromagnetic field produces a current in another wire called the receiver antenna. This current is then detected by a suitable device.

FIGURE 5.5

COMMUNICATION WITH ELECTROMAGNETIC WAVES

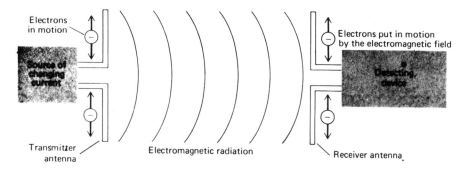

Charges moving in the transmitter antenna emit electromagnetic radiation which produces a current in the receiver antenna.

If the charges· that produce the electromagnetic wave are in a continuous back-and-forth oscillating motion, then the wave they produce will also oscillate continuously. Such a wave carries very little information. Its detection does indicate that someone is transmitting, but since the signal does not change with time, no additional information is obtained. The electromagnetic wave can, however, be altered at the transmitter to carry information. Such an alteration is called modulation. The radiation can be simply turned on and off, producing pulses in the form of the Morse code. These pulses can be detected at the receiving end and decoded. There are much more sophisticated modulation techniques with which sound and visual information can be transmitted. We will discuss these later.

At the time of Hertz there were still many technical difficulties in the way of a practical wireless. The main problem was in the method of detection. Detecting the electromagnetic radiation with sparks as Hertz did is obviously very inconvenient. In addition, the production of a spark requires a strong field. An electromagnetic wave, just as a sound wave, attenuates with distance from the source. Although the dissipative losses are very small, the wave spreads and therefore the energy in a given region decreases as the distance from the source increases. At distances of more than a few yards from the source, the field is too weak to produce a spark. During the following years many inventors concentrated on improving wireless communication. Prominent among them was Marchese Guglielmo Marconi. In 1898 Marconi and Captain H. B. Jackson, who had also worked with wireless communication, performed an impressive demonstration. During naval maneuvers they communicated over a distance of sixty miles. They received wide acclaim and the future of wireless communication was assured.

During the next three years Marconi kept extending the range of communication and by 1900 he was ready to try the great test, wireless communication across the Atlantic. Success here would firmly establish wireless as a most important communication technique.

When Marconi first proposed transatlantic wireless communication, there were many skeptics who believed that the project was impossible. They claimed that all electromagnetic radiation travels in a straight line and that therefore communication much beyond the horizon was not possible. But Marconi was confident in his project. He said, "These waves of mine will follow the earth." In fact both Marconi and his critics were right. Electromagnetic radiation does indeed travel in straight lines, but at high altitudes the earth is surrounded by a layer of ionized gas called the *ionosphere* which reflects radio signals back to earth. The ionosphere is produced primarily by the action of the short-wavelength components of the sun's radiation on the atoms in the upper atmosphere. This radiation produces ions by ripping electrons from the atoms. The ionosphere extends from thirty to two hundred miles above the earth and for radiation at radio frequency it behaves as a mirror by reflecting the incident radiation. As a result radio communication over long distances is made possible (Figure 5.6). At the time of Marconi's experiments nothing was known of the ionosphere and therefore the skepticism of his critics was well founded.

FIGURE 5.6

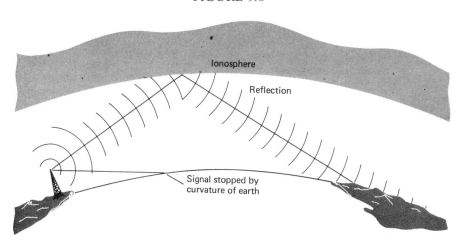

A radio signal propagated toward the ionosphere is reflected and can be detected beyond the line of sight of the antena.

After initial failures, Marconi finally succeeded in his transatlantic experiment. The electromagnetic signal was produced by sparks about four centimeters long and the current induced in the receiving antenna was detected by telephone. In describing this experiment Marconi wrote later:

"We reached Newfoundland on December 6 [1901] and erected our signal station on Signal Hill. On December 12 and in spite of a raging gale we flew a kite carrying an aerial some 400 feet long. About 12:30 in the afternoon a succession of three faint clicks corresponding to the prearranged signal sounded definitely and distinctly in the telephone held to my ear. This could only be that the electric waves sent out had traversed the 1700 miles of the Atlantic unimpeded."

Although Marconi's system was primitive and unreliable, the need for rapid communications was so great that it was·very quickly put into use in shipping and military communications.

The Marconi-type wireless communication system had many limitations. The generation of the carrier wave with sparks was cumbersome. The signals received were very weak and techniques for amplifying them did not exist. Until about 1906 all communication was with Morse code and voice communication seemed remote. To establish a practical radio voice-communication system new techniques had to be developed for the generation, amplification, and detection of electromagnetic signals.

A major breakthrough in radio technology came in 1904 with the development of thermionic vacuum tubes. That year the English physicist J. A. Fleming patented a two-element vacuum tube (diode). The vacuum diode was an excellent detector of electromagnetic radiation and is still widely used. Two years after Fleming's invention Lee De Forest inserted another electrode into the diode configuration and produced the triode, a device of tremendous importance. With the triode it became possible to amplify and conveniently generate radio signals. Now developments in communication became so rapid that by 1907 Reginald A. Fessenden broadcast speech over two hundred miles of the eastern coast of the United States. Through the work of hundreds of engineers and inventors, but notably De Forest, Armstrong, Pupin, Meissner, Poulsen, and Fessenden, radio technology was developed to its full potential. In 1920 the first scheduled broadcasting was begun in Pittsburgh by station KDKA. By 1923 there were more than five hundred transmitters operating in the United States.

Following the invention of vacuum tubes a whole new discipline and technology of electronic engineering evolved to utilize these devices. Electronic components were designed to be used in conjunction with the vacuum tubes in electrical circuits designed to perform the various functions required to transmit and detect radio signals.

Modern Radio

The simplest way to describe the modern radio system is in terms of the block diagram shown in Figure 5.7. Each block in the diagram represents a necessary function in broadcasting and receiving.

At the radio station the microphone converts the acoustical sound waves into an electric signal which is called the audio signal. The audio signal is then amplified, and its information content is imposed on a high-frequency radio signal, which is radiated into space. Why do we not

FIGURE 5.7

BLOCK DIAGRAM OF A RADIO SYSTEM FROM MICROPHONE TO SPEAKER

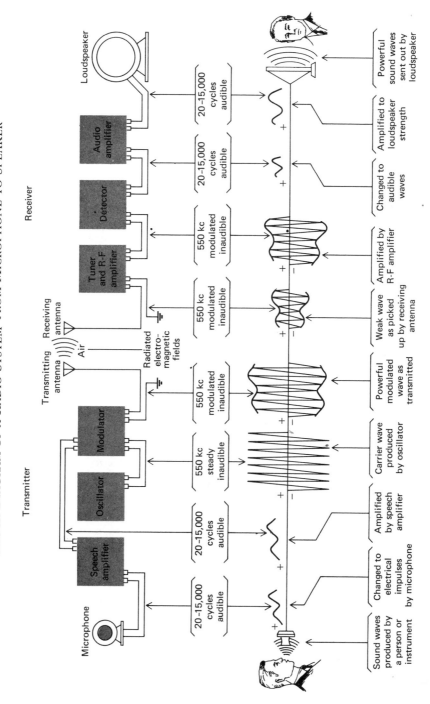

broadcast the audio signal directly? It is, after all, an electric signal and therefore could be radiated into space. This, in fact, could be done; but there are several reasons for not doing it. As mentioned earlier, the longer the wavelength of radiation, the larger is the antenna required to launch the radiation efficiently into space. The frequency of the audio signal is, of course, the same as the frequency of the sound wave. This frequency is low—between twenty and a few thousand cycles per second. In order to efficiently launch electromagnetic radiation in this frequency range, the transmitting antenna would have to be inordinately large. Furthermore, if broadcasting were done directly at audio frequencies, all broadcasters would be forced to transmit within the same frequency range, which would result in intolerable confusion. For these reasons broadcasting has to be done at frequencies considerably higher than the audio range. Thus the information contained in the audio signal has to be superimposed onto a high-frequency radio signal, which is then radiated into space. The high-frequency radio signal is called the carrier and the process by means of which the audio information is placed on the carrier is called modulation. The frequency of the carrier is typically above 500,000 cycles per second. In order to avoid interferences each station in a given broadcasting region uses a somewhat different carrier frequency.

We have already discussed one very simple modulation technique, the transmission of the Morse code. Here the carrier is simply turned on and off, transmitting pulses that convey the alphabet. The information is contained in the duration of the radio frequency pulse. A more sophisticated modulation technique has to be used if the full audio information is to be implanted onto the carrier.

The carrier frequency alone contains no information. In order to put the information contained in the audio wave onto the carrier, some feature of the carrier has to be varied in accord with the audio signal. Because there are two parameters that describe an electromagnetic wave, the amplitude and the frequency, we immediately have two ways of implanting the information onto the carrier. We can vary either the amplitude or the frequency of the carrier in accord with the audio signal. If the audio information is contained in the amplitude of the carrier, the modulation is called amplitude modulation (AM). If the information is carried in the frequency of the carrier, the modulation is called frequency modulation (FM). At present both types of modulation are used. The block diagram in Figure 5.7 illustrates amplitude modulation. Returning to our block diagram, we can see that the carrier is generated by an oscillator. In the modulator, the amplitude of the carrier signal is varied in accord with the audio signal to produce the modulated carrier. The amplitude of the modulated carrier has been outlined to point out that it follows the shape of the audio signal. The modulated signal is then radiated into space by the antenna. At the receiving station the electromagnetic radiation causes current to flow in the antenna.

Our space is filled with electromagnetic radiation at various carrier frequencies emitted by different broadcasting stations. Currents produced by all these carrier frequencies flow in the antenna and they have to be sorted out in some way. It is the tuner that picks out only that carrier frequency in which the listener is interested. The amplitude-modulated carrier is then amplified by the radio frequency (RF) amplifier and is applied to the detector which extracts the audio signal. The audio signal is then amplified and applied to a speaker, where it is converted to an acoustical signal.

Although amplitude modulation is the simplest method of modulation, it has a major fault. Most natural and man-made electrical disturbances can impress themselves on the carrier wave and produce amplitude variations on it. At the receiving station these unwanted amplitude variations are detected and produce noise. For example, a flash of lightning produces a pulse of electromagnetic radiation that passes through the receiver as a crackle. Electric motors, generators, and transformers induce similar disturbances. As we have previously mentioned, there is another way of modulating a carrier wave and this is by varying its frequency in accord with the audio signal. Frequency modulation removes the system from the domain of these noise sources. In frequency modulation the frequency of the carrier is varied by an amount proportion to the amplitude of the sound. The rate at which this variation takes place is determined by the frequency of the sound wave.

5.6 TELEVISION

To those who have been reared in a technological environment television may seem no more mysterious than radio or any of the other devices that permeate our lives. But to people who first encounter technology, television is most perplexing. It is not unusual to hear a disembodied sound, but to see changing images without an obvious source is truly puzzling. Television is, in fact, a much more complicated device than radio even though the basic ideas are relatively simple.

A direct visual image is conveyed by light which is itself high-frequency electromagnetic radiation. However, the propagation properties of light and human optics are such that it is not possible for us to detect a detailed image from distances further than a few hundred yards. In order to transmit the visual image over long distances, the image information is converted into an electric signal which modulates a high-frequency carrier. The modulated carrier is then radiated into space, picked up by the receiving antenna, and demodulated. The resulting signal is reconverted into a visual image. The process is therefore similar to radio broadcasting. The additional complexities arise in the conversion of the visual images into electric signals and vice versa.

The basic ideas of modern television broadcasting were first published in 1908 by A. A. Campbell-Swinton in *Nature* magazine. However, it took more than twenty years before the technology was sufficiently developed to build a working system. The key to image transmission is the conversion of the visual image into an electric video signal. This is done by a technique known as *scanning*. The visual image is examined or scanned line by line with a narrow beam of electrons. As the electron beam strikes each point of the image an electric current proportional to the local image brightness is generated. This current is called the video signal. In the commercial television systems used in North America the electron beam scans the whole screen in 525 lines.

About thirty full pictures per second must be displayed in order for the eye to perceive the sequence as continuous motion. Thus to convey motion the full scanning is repeated thirty times a second. A portion of the video signal that would result from scanning the letter *A* is shown in Figure 5.8. We note that a horizontal synchronization pulse is added at the end of each scan. This pulse controls the image reproduction at the receiving end. The video signal is used to amplitude-modulate a carrier in a way that is identical to the method we discussed in connection with amplitude-modulation broadcasting. The carrier frequency is in the neighborhood of 50 megacycles, but each television transmitting station in a given locality uses a different carrier frequency. The carrier also contains the audio information associated with the television picture, which is usually transmitted by frequency modulation. The modulated carrier containing the audio and video signals is launched into space by the transmitter antenna and is picked up by the receiving antenna. After demodulation the audio signal is applied to a speaker, and the video signal is converted into an image by a device called the *cathode-ray tube* or, simply, *picture tube*.

FIGURE 5.8

SCANNING

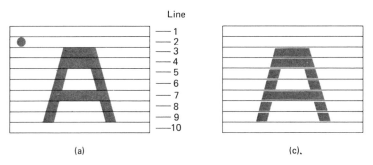

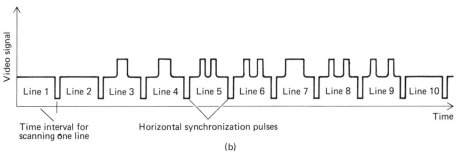

(*a*) The letter *A* scanned by ten lines. (*b*) The resulting video signal (the horizontal synchronization pulses are explained in the text). (*c*) The reproduced image.

The origins of the cathode-ray tube can be traced back to the 1850s (Figure 5.9). It consists of an electron gun that produces a beam of electrons, and a screen coated with a fluorescent material that glows at the point at which it is struck by an electron beam. The electron beam again scans the fluorescent screen line by line in synchronization with the scanning pattern at the transmitter. The motion of the beam in the cathode-ray tube is controlled by the synchronizing pulses that are sent along with the video signal. The amount of light emitted by the fluorescent screen is proportional to the number of electrons hitting it. The number of electrons in the beam is controlled by the video signal that is applied to the intensity control. A large video signal allows more electrons to pass through to the screen than does a small signal. In this way the image intensity of the transmitted picture is reproduced line by line on the receiving screen. Although the picture is built up line by line, the

FIGURE 5.9

PICTURE TUBE

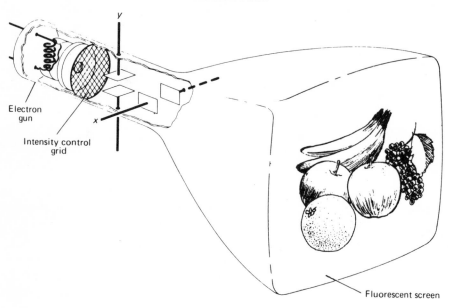

Electron gun

Intensity control grid

Fluorescent screen

scanning is sufficiently rapid that the eye responds only to the built-up image as a whole.

Color Television

Color television transmission began in the United States in 1954. It is based on the principle that any color can be reproduced by the proper combination of the three primary colors red, green, and blue. By means of mirrors and filters the three color components of the image are separated and each color component is projected on a separate television camera which produces a video signal proportional to the respective color content of the picture. At the receiver the television tube contains three electron guns, each of which is controlled by the video signal of one of the primary color patterns. The screen of the receiving tube is composed of three separate sets of phosphor patterns which produce, respectively, the three primary colors when bombarded by the electron beam (Figure 5.10). The phosphor patterns are uniformly distributed and the electron beam of each gun is so focused that it hits the phosphor corresponding to the color that controls it.

FIGURE 5.10

THE TRICOLOR TELEVISION PICTURE TUBE

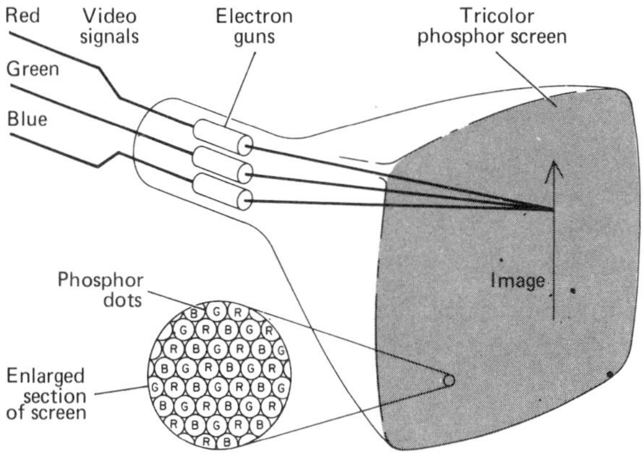

5.7 RECENT DEVELOPMENTS AND FUTURE TRENDS

One of the most important technological events of the 1950s was the development of semiconductor devices such as transistors and semiconductor diodes. These solid-state devices perform the same functions as the conventional vacuum tubes, but they are so much smaller that they have made feasible completely new ventures. Semiconducting devices are not only smaller but they also require much less operating power. This has given rise to new concepts in battery-operated appliances, such as radios, hearing aids, and electric watches. Because semiconductor devices to not have heater filaments, they are ready for instantaneous operation.

During the last fifteen years diffusion and photochemical techniques have been developed that make it possible to produce on a single small semiconductor crystal complete circuits consisting of hundreds of individual components. These are called integrated circuits. Transistors, diodes, resistors, and capacitors can all be produced on the crystal by forming microscopically small regions with the desired properties. The small size of integrated circuits has profoundly changed electronic design concepts and feasibilities. This is clearly illustrated in computer technology. The early IBM computers using vacuum tubes could perform 39,000 additions in one second. The introduction of transistors in 1955 increased the computer capability to 204,000 additions per second. With the newest integrated circuits the computer can perform 15 million additions in one second. Semiconductor electronics has also been one of the cornerstones of the space program, in which miniaturization is absolutely essential.

The miniaturization trend will continue. There is, however, at least on one level, a limit on the smallness of the device. The device must remain large enough to be manipulated. Recently we had a rather sophisticated amplifier constructed, using the most advanced solid-state components. Had conventional vacuum tubes been used, the components would have occupied a volume of about two cubic feet. The total electronics were about the size of a dime. However, the amplifier had to have a number of switches and dials to control its operation. The smallness of these components is fundamentally limited by the size of the hand. These switches and dials had to be mounted on a panel that is part of the device. In the end this amplifier was not much smaller than an amplifier with conventional electronic components. It was, however, less expensive and lighter.

Integrated circuits have also reduced the price of communication devices. Complex electronic equipment can now be constructed at a relatively low cost. Citizen-band radio with which a person can both receive and transmit radio messages is now within the reach of many people. One can envision a future in which every person is in continuous two-way contact with the whole world. A small personalized transmitter-receiver will link us all even in the deepest wilderness. Whether such an intense communication network is desirable is another question.

In order to transmit information a certain amount of space is required in the frequency spectrum. The larger the information content of the signal the greater is the frequency space required to transmit it. The transmission of a television signal, for example, requires nearly a thousand times the space needed for the transmission of sound. The higher the frequency of the electromagnetic carrier wave the greater is the amount of information it can transmit. For this reason the frequency of carrier waves used in communication has been getting progressively higher. At present many intercity and intercontinental communication links are made with carriers in the frequency range of kilomegacycles (10^9 cycles) per second. Radiation in this frequency range is called *microwave* radiation or simply microwaves.

Following World War II, microwave technology was applied to communication. Intricate microwave networks have been built in North America and Europe. The networks consist of microwave towers placed at intervals of twenty-five to thirty miles. An antenna on top of the tower receives the signal, which is amplified and transmitted toward the next tower. The antenna is usually horn-shaped or parabolic in order to focus the radiation onto the detector. At present most microwave transmission consists of telephone and teletype messages and television programs being transmitted from a central source to the local stations. Usually the many individual signals are simultaneously transmitted on a single carrier, using multiplexing techniques.

An intercontinental microwave link was provided by space technology. The space program required techniques of communicating with satellites and space probes. These techniques were very quickly incorporated into the communication system. On August 12, 1960, the first communication satellite, Echo I, was launched. Echo I was simply a large balloon, thirty meters in diameter, which reflected microwave radiation back to earth. The radiation reflected back to earth spread into a large sphere and since it was not amplified, the signal detected by the receiving antenna on the earth was very weak. The Echo satellites did, however, demonstrate the feasibiity of this type of communication.

The communication satellites following the Echo series have all been active. These satellites contain amplifiers to strengthen the signal and antennas to focus the signal preferentially back toward earth. The power for the amplifiers is obtained from banks of solar cells (Figure 5.11).

FIGURE 5.11

MICROWAVE COMMUNICATION BY SATELLITE

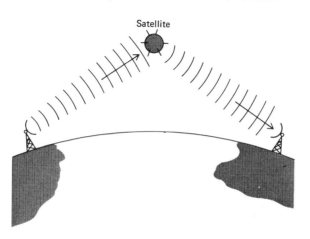

Dozens of satellites have already been launched into different orbits, some only a few hundred miles above the earth, others many thousands of miles into space. Almost from the beginning the communication satellites worked well and today transcontinental television, radio, and telephone transmission via satellite is routine.

The trend towards higher carrier frequencies will continue. With the invention of *lasers* it has become possible to generate carrier frequencies 100,000 times higher than those currently in use. The information-carrying capacity is of course correspondingly higher. It is likely that lasers will be used as part of the future communication network.

A fascinating possibility in the future is communication with extra-terrestrial civilizations. Some estimates indicate that in our galaxy alone there may be as many as two billion stars with planets some of which are suitable for intelligent life. With the modern radiotelescopes, the possibility of detecting extraterrestrial artificial radio signals is no longer in the realm of science fiction.

Predictions about technology usually tend to be conservative. When the wireless telegraph was first put into operation, very few people could foresee the development of our television system out of that primitive technology. In 1957 when the first Sputnik orbited the earth, the landing on the moon still seemed in the very distant future. The exact course of our future technology is not totally predictable. One thing is certain, however: the technology will continue to grow. Whether this technology will improve or degrade the quality of our lives depends on how we use it. If the past is any indicator of the future, it will do both.

5.8 COMMUNICATION AND SOCIETY

Modern communication technology has had a great impact on our society. Politics, merchandising, and law enforcement, to name a few areas, have all been altered by communication technology. Television, which in the average American home is switched on six hours a day, has had the most evident effect. Political campaigns are routinely waged on TV screens and often the abilities of the candidates to use the medium determine the outcome of elections. Advertising campaigns sway public consumption patterns, often in irrational ways. The viewers are urged to buy junk food or big cars and they respond. Television has converted sports into a big business and made us a society of spectators.

The intense involvement of people with television has been a concern now for many years. Important questions have been raised in this connection. A large number of TV programs are violent and deal with crime. Is the behavior of people affected by watching so much violence and chaos? Children are among the heaviest users of television. The hours they spend in front of the TV set day after day are likely to have an effect on their development.

In May 1978, Chancellor Helmut Schmidt of West Germany suggested that West Germans have one television-free day each week. "We are not talking enough with each other. That goes for married couples, parents, children and friends. I am disturbed by the fact that we are becoming more and more tongue-tied." He claimed that television had many positive aspects but often gave people a false picture of real life and tended to portray violence as a normal occurrence.

It is tempting to focus entirely on the adverse effects of television. There is, however, another side to this medium. Even at its worst TV has relieved a lot of loneliness in ways that are safer than drugs and alcohol. There have been many studies that have examined the effect of television on these and other areas of societal concern, and numerous journal articles and books have been written on this subject. While there is no unanimous agreement on the various issues, our awareness of the problems has increased. As a result television programming is getting better and more thoughtful. Some excellent programs are now being presented by public and at times also by private networks. These programs, while superbly entertaining, have also helped us to understand science, music, literature, social problems, and many other aspects of society. Recently the Public Television Network broadcast the opera *La Boheme.* The audience for this single performance was greater than the combined audiences for all the past performances of this opera.

While television has received the most attention, other areas of communication technology have also had a major impact on societal activities. Computers have provided the means to store a large amount of data which can be easily retrieved and distributed. In most cases the use of this data is legitimate and productive. Telephone-computer links, for example, are now routinely used by banks and other businesses to check the financial status of customers and by airlines to schedule ticket sales. There is, however, a concern about the misuse of these systems. All too often confidential data are retrieved from the computer in ways that violate the rights of the individual to privacy.

Although there are numerous examples to justify the use of communication technology in law enforcement, some of the more recent applications of technology to crime detection have been challenged. During the past few years electronic eavesdropping devices have become so sophisticated that it is now possible to monitor the activities of people with almost complete secrecy. Their use has raised a whole range of legal issues which are still not completely resolved. The problem centers on defining the narrow borderline between crime prevention and the violation of the civil rights of individuals.

The social effects of communication technology deserve a more thorough coverage than we were able to provide in this chapter. Some of the issues we have not discussed are raised in the "Exercises" section. One important point emerges even from a superficial examination of these issues: communication technology, like all other human endeavors, can be used both to enrich society and to subvert it. Television can transmit a concert as faithfully as it transmits a deodorant commercial. The way this or any other technology is used depends on society.

5.9 EXERCISES

1. Below are listed some of the firsts in communication. Examine the initial impact of these innovations and comment on the prevalent opinions about them at the time of their introduction (starting point: newspaper of the time).
   a. The first intercity telegraph communication in the United States occurred on May 24, 1844.
   b. Bell's first telephone conversation took place on March 10, 1876.
   c. On July 27, 1866, the transatlantic telegraph cable was completed.
   d. On December 12, 1901, Marconi received the first wireless transatlantic message.
   e. The first scheduled radio broadcast was on November 2, 1920.
   f. The first regular television service was opened in London on November 2, 1935. In the United States TV broadcasting began on July 1, 1941.

2. It is not unusual for a meek and shy person to become aggressive on the telephone. This shows the drastic change that the telephone can cause in interpersonal relations. Examine this aspect of the telephone.

3. The telephone has become an important tool of governing. Its effective use allows for great centralization of power. Examine this role of the telephone in industry and (or) government.

4. People are devoted to specific types of TV programs. Is there any correlation between programs and the type of people who watch them?

5. Discuss the effect of adult TV programs on child audiences.

6. Examine the use of radio as a propaganda tool during World War II.

7. Examine the effect of mass media on international conflicts.

8. In the United States radio and television advertising compete for different markets. The radio stations are oriented towards local interests while the television companies are national. Examine this proposition in greater detail.

9. Compare TV and radio programs in different parts of the world, and relate them to the culture and socioeconomic conditions of the countries.

10. Examine the effect of mass media on political life in the United States.

11. The Civil War in the United States was the first major conflict in which modern technology—trains, shipping, and telegraph—played a significant role. Examine the effect of these on the course of the war.

12. Examine the present use and the future potential of TV as an educational tool.

13. Examine the technology and legal aspects of eavesdropping.

5.10 BIBLIOGRAPHY

1. Ashford, T. A. *The Physical Sciences.* New York: Holt, Rinehart & Winston, 1967.
2. Berkowitz, L. "The Effects of Observing Violence." *Scientific American* 210, 1964.
3. Brenton, M. *The Privacy Invaders.* New York: Coward-McCann, 1964.
4. Buzzi, G. *Advertising: Its Cultural and Political Effects.* Minneapolis: University of Minnesota Press, 1968.
5. Chester, E. W. *Radio. Television and American Politics.* New York: Sheed & Ward, 1969.
6. Davidovits, P. *Communication.* New York: Holt, Rinehart & Winston, 1972.
7. Ford, K. W. *Basic Physics.* Waltham, Mass.: Blaisdell, 1968.
8. Martin. J. *Future Developments in Telecommunication.* Englewood Cliffs, N.J.: Prentice-Hall, 1971.
9. McLuhan, M. *Understanding Media: The Extensions of Man.* New York: Signet, 1964.

Chapter Six

COMPUTERS

6.1 THE COMPUTER REVOLUTION

The Computer Revolution refers to a vast complex of changes occurring in the handling of information and the processing of data, words, and ideas that have occurred during the twentieth century, mainly since World War II ended in 1945. These changes have caused a great many operations of arithmetic, logic, and problem-solving to be performed by machines a million times faster than they used to be performed by human beings.

Usually an economic revolution occurs when some important economic process becomes ten times faster or one-tenth as expensive. But in the case of the Computer Revolution a great many processes have become from ten thousand to a million times faster over the last twenty-odd years.

These changes are sometimes called the Electronics Revolution and sometimes called the Second Industrial Revolution. But the term we shall chiefly use is the Computer Revolution.

It is not enough for experts alone to understand the Second Industrial Revolution. The public as a whole also needs to understand it, because it affects them intimately in very many ways.

As is well known, the first Industrial Revolution resulted from the introduction of power-driven machinery to replace hand labor, and began essentially in England about 1760–1800. This revolution drastically altered the lives of human beings, their methods of working, and their habits of living. In addition, the Industrial Revolution brought about radical changes in cities, factories, political organizations, government, business, and society as a whole.

Although the changes that occurred as a result of this first revolution were great, the changes caused by the Computer Revolution are becoming even more pervasive. This revolution has altered and will alter many occupations of white-collar workers, blue-collar workers, engineers, draftsmen, technicians, teachers, mathematicians, and many more workers besides. The changes mainly result from transferring to machines immense loads of (1) clerical record keeping and computation; (2) watching and monitoring processes in factories, refineries, etc.; (3) engineering and technical calculations; (4) looking up information in business and government files and making decisions as a result of data; (5) calculating solutions to problems instead of solving them by the "seat of the pants" or "armchair" method; etc.

6.2 A BRIEF HISTORY OF CALCULATING DEVICES

Men have been seeking to count, to measure, and to know more about every aspect of life for at least twelve thousand years. By measuring, counting, charting, and graphing events in time and regions of land, men have gradually gained some understanding and some control over parts of the environment.

The first calculating systems appear to have been pebbles, notched sticks, and knotted strings (see Figure 6.1). Later, the "abacus"—a slab with pebbles and still later a frame with beads—was invented for calculating (Figure 6.2). Medieval Europeans used "counting boards" resembling checkerboards for calculating aspects of trade and business. The term "counting house" originated in Italy in the 1300s. The words "carry" and "borrow" came from moving disks on the counting boards. Also, many board games, such as checkers, backgammon, etc., derive from the counting boards.

Gradually, more advanced machines for counting and calculating were invented. In 1649 Blaise Pascal, a French philosopher and mathematician, who was working as a tax clerk, invented a counting machine called an adder (Figure 6.3). Gottfried Wilhelm von Leibniz, a German mathematician, advanced mechanical counting systems with an invention in 1694, using the principle of repeated addition. These are just a few examples of a long series of inventions that became continually more complex and more useful for calculating.

Charles Babbage, an English mathematician and actuary, developed some new basic concepts for a calculating system that he called the

FIGURE 6.1
KNOTTED STRINGS

FIGURE 6.2
AN ABACUS
WITH A FRAME

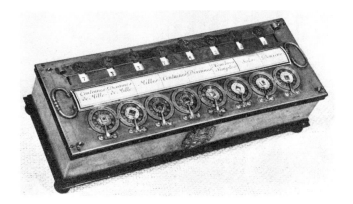

FIGURE 6.3
PASCAL'S ADDER

From an original in the IBM Corporation Antique Calculator Collection.

Difference Engine. This machine was begun in 1823 and was planned in order to calculate numerical tables. The machine would store numbers in a series and add sets of differences successively in order to compute the next number in a series. Such tables include navigation tables so that ships at sea could determine their positions. In spite of twenty years of monetary support to Babbage from the British government, however, the Difference Engine was not finished (see Figure 6.4). The essential reason was an insufficient level of mechanical engineering to express in working hardware Babbage's concepts, which were in fact completely modern concepts of a computer.

About 1802, a French inventor, Joseph Jacquard, contrived a machine for automatic weaving. It used successive cards punched with successive patterns of holes in order to position the threads of a loom during the weaving process. The cards for the Jacquard loom were the forerunners of the common punched cards in use today.

FIGURE 6.4

BABBAGE DIFFERENCE ENGINE

From a replica in the IBM Corporation Antique Calculator Collection.

6.3 FOUR ESSENTIAL DEFINITIONS

In order to discuss adequately the far-reaching social consequences of the Computer Revolution, we need first to answer several important questions in regard to computers:
1. What is a computer?
2. What is an algorithm?
3. What is a computer program?
4. What is a programming language?

The essential meanings of these four terms are simple, and are even illustrated in the common ordinary everyday world. But these terms have received some special meaning in the realm of computers. Knowledge of the special meanings helps to remove the mystery and consequent fear of the unknown related to computers.

A Computer

What is a computer?

The word "computer" derives from the Latin "computare" meaning "to reckon or calculate"; and that in turn is derived from "putare" meaning "to think," and the Latin prefix "cum," meaning in this case "intensely," i.e., "to think intensely." It is rather a nice derivation.

The dictionary gives definitions of "computing" as follows:
1. The process of numbering, reckoning, or estimating.
2. The process by which different sums, numbers, or items are compared.
3. Numbering or counting or determining by reckoning; ascertaining the aggregate or total result.

To these should be added a definition based on what is actually happening inside the machines:
4. Handling information (data, words, or ideas) by long sequences of arithmetical and logical processes under a flexible control.

Many definitions of "computer" can be found in the literature. Like the six blind men reporting on an elephant and each of them talking about a different part of the elephant's anatomy, the definitions of a computer express many discordant perceptions and interpretations of a computer.

A definition of a "computer" furnished in a glossary by the IBM Corporation is: "A computer is an information-processing machine." Here "information" may be numbers or alphabetical characters, words, mathematical symbols, etc. Unfortunately, this does not distinguish a computer from an adding machine or a clock: both process information and neither is a computer.

Another definition is: a computer is a machine that is able to take in and store information (problems, numbers, instructions), and to perform automatically a changeable sequence of reasonable operations on the information, as may be required for solving a problem and to put out the answers.

This definition is better and does exclude the adding machine and the clock. It also does not necessarily invoke electronics, although the vast population of computers would never have come into existence unless electronics had given them enormous speeds and consequently enormous power.

The above definition includes a human being. And in fact, of course, a human being is a computer. The original meaning of "computer," in the 1940s and earlier, was always a person, never a machine. For example, during World War II, about a hundred persons (college graduates and trained Army employees) were employed at the Moore School of Electrical Engineering at the University of Pennsylvania to compute, using desk calculating machines, the trajectories for each type of artillery shell with different values of muzzle velocity and different angles of fire. Each ballistic table required the computing of hundreds of these trajectories. In twenty hours one person could compute the trajectory of a shell which was in the air for sixty seconds. By contrast, one of the first electronic computers, the ENIAC, could perform this same task electronically in thirty seconds.

The main difference between a person computing and a machine computing is that the machine performs calculating tasks with very high speed, much more accuracy, and far more reliability than is ever possible for a human being. A modern computer can accomplish in ten minutes a problem-solving task that would take a human being an entire lifetime. So the computer is like the fabulous "Fountain of Youth" of Ponce de Leon, giving a human being more than ten thousand lifetimes for calculating answers to his questions. And man is the most curious of all animals.

The speed, precision, accuracy, and reliability of computers in problem-solving and information processing has made people wonder whether or not "machines think." What is thinking? Can a machine do it? If "thinking" is defined as an objective process, the answer is "yes"; a machine can do it. If "thinking" is defined as what only a human being can do, or as a vague, subjective process in terms of other vague words, then a machine cannot do it.

An Algorithm

The second critical term that needs to be understood in connection with computers is "algorithm." An *algorithm* is an effective calculating rule. More specifically, it is a complete sequence of specific steps in arithmetic or logic or both which proceeds from data (or input information) to answers (or output information). In others words, an algorithm is:

— an exact recipe;
— a set of consecutive instructions that works;
— a procedure, as in an office of laboratory or factory, for taking in certain given information and putting out certain desired information.

An interesting article, "Algorithms," by Donald E. Knuth, a professor of computer science at Stanford, in *Scientific American* for April 1977, begins:

> Ten years ago the word "algorithm" was unknown to most educated people: indeed it was scarcely necessary. The rapid rise of computer science, which has the study of algorithms for its focal point, has changed all that; the word is now essential. There are several other words that almost or not quite capture the concept that is needed: procedure, recipe, process, routine, method, rigmarole. Like these things an algorithm is a set of rules or directions for getting a specific output from a specific input. The distinguishing feature of an algorithm is that all vagueness must be eliminated; the rules must describe operations that are so simple and well defined that they can be executed by a machine. Furthermore an algorithm must always terminate after a finite number of steps.

The main difference between an algorithm and a "program" (which will be defined in the next section) is that (1) a program is expressed in instruction codes that a computer can accept (or "understand"), while (2) an algorithm is a plan or pattern or rule for calculation irrespective of coded instructions.

The word "algorithm" comes from the name of a Persian scholar, al-Khwarizmi, of about 825 A.D., whose textbook on arithmetic had a wide influence for many centuries.

A Program

The third critically important idea in connections with computers is a program. A program is a sequence of computer instructions that solves a type of problem. In other words, a program is an algorithm expressed in the language of a particular computer.

Many people, perhaps a majority, make no distinction between "algorithm" and "program." This is natural, because for more than twenty years in the growth of the computer field, no distinction was actually made. Recently, however, the distinction has been more and more observed. The reason is that the pattern of a program which is an algorithm requires the separation of concepts to refer to them clearly.

What is a reasonable estimate for the total number of programs residing in computer installations?

There are probably about two hundred thousand computer installations, and probably about a thousand programs on the average at each installation. This estimate leads to approximately two hundred million programs for producing answers from data, answers people are interested in, answers they will pay money for, and answers that affect human decisions.

Obviously, there is enormous duplication in purposes of programs. But different computers require different forms of instruction, different installations require different degrees of exactness, and so on.

A Programming Language

We come now to the fourth definition, "programming language."

A programming language is regularly an artificial language that has been constructed to express instructions to a machine in such a form that the machine can accept them.

One example of a programming language is a set of 1's and 0's corresponding to punched holes and no punched holes, or switches on and switches off. The language is composed in such a way that each set of signs gives information to a machine, and enables it to go from the current state to the next state. For example, the symbol 11 000 001 when punched in a paper tape can cause a teletype to print the letter A. The symbol 11 000 010 can cause printing of the letter B, and so on. This language is usually called Binary Notation or Machine Language.

A second example of a programming language is called Assembly Language. In this case the language can be translated by a computer program called an "assembler" peculiar to the type of machine. Such instructions as the following are in the assembly language of a kind of machine called the PDP 9 made by Digital Equipment Corporation:

200/	LAC	100
201/	ADD	101
202/	DAC	102
203/	JMP	16000

What does this mean?

In location 200 put the instruction "load the accumulator (LAC) with the contents of location 100."

In location 201 put the instruction "ADD 101," which means "add the contents of 101 to the accumulator."

In location 202 put the instruction "deposit the contents of the accumulator (DAC) in location 102."

In location 203 put the instruction "JMP 16000," which reports "the next instruction is found in location 16000" and tells the computer to "transfer control to location 16000."

How do we operate this program?

1. Put a first number that we want to add in location 100: say 12.
2. Put a second number that we want to add to it in location 101: say 23.
3. Tell the computer to start the program at location 200. For this purpose we type in to the computer's supervisory program (called DDT) the command 200', where the quote mark is a special typewriter symbol meaning "go" and 200 tells DDT the starting location.
4. Then we look in location 102, and find the resulting sum, which is 35.

The symbols LAC, ADD, DAC, JMP, and about forty others, together with a small number of conventions, together constitute what is called the Assembly Language for this type of computer.

The two kinds of programming languages we have so far mentioned, Binary Notation and Assembly Language, are very useful for particular machines. But it is necessary to have programming languages that are independent of particular computing machines. They can be used to describe algorithms so that they can be understood on a great many different machines.

Two main kinds of machine-independent programming languages are: general purpose, and special purpose. Special-purpose programming languages have evolved for civil engineering, small businesses, text editing, graphic descriptions, and many other special purposes. General-purpose programming languages enable the procedures for solving almost all problems to be described.

There are many general-purpose languages: FORTRAN, COBOL, ALGOL, BASIC, PL/1, etc. FORTRAN (so called from "*FOR*mula *TRAN*slation") was worked out about 1956, and has been used for a vast number of problems that come from scientific and engineering fields. COBOL (for *CO*mmon *B*usiness-*O*riented *L*anguage) was worked out about 1962, and has been used to specify procedures for a vast number of business problems. BASIC was worked out at Dartmouth College, Hanover, New Hampshire, about 1967. It has been widely used for students to solve problems that arise in their courses and is implementd with time-shared interactive computer systems. Also, with extensions for dealing with files, BASIC has been used for a great many problems of small businesses. ALGOL (*ALGO*rithmic *L*anguage) was worked out by European and United States programming specialists in Zurich, Switzerland, in 1960. PL/1 is a product of IBM from the mid-sixties which contains many fine features of FORTRAN, COBOL, ALGOL, and Assembly Languages.

In summary, present-day civilization in the advanced countries is complex, engineering-wise and technologically, and complex business-wise and industrially. Such a civilization produces an enormous growth in the information to be handled and processed.

It is clear that this is the force that provided the energy and the urgency behind the great development of automatic handling of information, expressed in computing and data-processing systems, the Computer Revolution.

As a result, the number in use of ordinary computers, minicomputers, and now since 1975 microcomputers, is well over a million. The average number of calculating and reasoning operations per second in the world is well over one hundred billion. So we can realize that a great many of the systems, processes, and controls in the modern world would quickly and completely stop if computer power were suddenly rendered non-existent.

6.4 COMPUTER APPLICATIONS

In applications use, we are thinking generally of a series of programs that will allow people to perform a series (or choice) of tasks in a given area. These groups of programs are called *software packages*. They are written by programming specialists for different purposes and disciplines. Some applications packages are *discipline-oriented*—they are aimed at specific disciplines, which might be engineering, mathematics, physics, art, poetry, music, architecture, medicine, business, political science, geography, etc. Other applications packages are *task-oriented*—they are written to accomplish certain jobs or tasks for many kinds of people, but the focus is on the task: using the statistical package on the computer library; drawing different types of maps; achieving a variety of charts and graphs from data; or executing a series of engineering drawings for the erection of a building.

Many applications packages are written in such a way that the user merely calls the routines (subroutines) from the computer library, adding personal data. The persons using these systems do not have to know a great deal about programming language and computer science. Such systems are called *user-oriented systems*.

In using the computer for applications work, different types of input and output devices or machines are used, called *peripheral devices*. They are located outside the computer, and are connected (or interfaced) to it. A few examples of peripheral output devices are:

—Automatic drafting machines (or plotters) that draw maps, charts, graphs, musical notation, buildings, machine parts, art graphics;

—Automatic milling machines that cut and machine parts from wood, plexiglas, or metal;

—Microfilm plotters that produce 35 mm film (microfilm) for reports of many types, with or without pictures or graphs.

Varying Levels of Involvement in Computer Applications

In many applications, the computer is an "aid" to performing routine tasks for specific purposes—it is not a "take-over" of the application of discipline. Rather it augments, enhances, and speeds up the work of human beings, relieving them from tedious manual drudgery.

Using the computer does not necessarily mean that one becomes a computer professional. There are many different levels of computer use, each requiring specific levels of computer knowledge. Here are some examples of quite different levels of computer applications involvement and their requirements:

Level A—In two hours a bank clerk learns how to input and output data for special purposes, which may be loans, deposits, interest, monthly or quarterly reports. An *interactive terminal* system is frequently used in this type of application, and it requires no previous programming background. The terminal resembles a typewriter keyboard with paper output, and the clerk "interacts" or communicates directly with the programs that are being used. The user may ask questions and receive help from the computer system. The terminal may also consist of a display screen, resembling a television tube, called a cathode ray tube (or CRT), on which output appears. This type of interactive use is much more interesting for the clerk than the previous manual-entry method, and more work is accomplished during the same amount of time with fewer errors.

Level B—A government assistant is enrolled in a workshop and learns how to input data and output many different types of charts and graphs that formerly were drawn by hand. This is a special graphing/charting system, designed for non-programmers. The assistant is enrolled in a training session for two hours a day, once a week, for six weeks, on company time. At the end of the six-week period, the assistant can quickly achieve a wide range of charts and graphs required for many types of reports. The system is easy to use, and is far more interesting than the tedious manual methods used before. Instead of graphing the data in one form, the same data can readily be presented in several different ways, enabling human beings to understand more fully the relationships and implications of the processed data. Often, because the assistant is able to perform more work in less time, a promotion in position and salary is made. Again, from computer applications, a human being is freed from a demanding and very tedious task.

Level C—A geographer enrolls in a three-unit upper division course sponsored by his department in the university, in order to learn how to make different types of computer-aided maps, using a special systems package designed for geographers. It requires no previous computer programming experience. At the end of the course, the geographer can achieve a wide variety of maps quickly and precisely. In addition, he has learned not merely how to use the mapping system, but has also learned how to design and input *new* maps that may be needed. The geographer does not yet know how to write new programs and subroutines, but can vary and depart from the initially taught routines. These maps can be drawn in ten to twenty minutes instead of requiring forty hours of manual drawing time. Further, the computer-aided maps are more accurate, and are even beautiful (as are the charts and graphs achieved in Level B above). The geographer is freed to think and do more complex tasks.

Level D—A college instructor enrolls in a weekend seminar sponsored by the National Science Foundation, or a regional computer consortium, etc. During this brief period, the instructor receives in-depth help in learning a discipline or task-oriented system useful in teaching and/or research. By the end of the weekend seminar, the instructor has developed sufficient skill to use the new system, and is prepared for further computer instruction in specific applications areas. This person will continue studying computer applications in a variety of ways: a second workshop, a classroom situation, or a seminar offered by the local computer center. He or she will become increasingly proficient in a range of computer-aided tasks that are useful in teaching and research. This new applications knowledge will be used in teaching, and many students will learn how to use specific techniques and computer applications in their own discipline. This level of use is a continuing one that grows in complexity, for the instructor will often write programs for the students to use. The teaching of computer applications in this instructor's courses will change and enhance each class—the teacher will be able to teach more, and the students will learn more and do more.

Level E—A computer science major is intrigued by business applications, and decides to add a business minor to his current studies, to learn how to use and to understand business systems applications. This requires more training than can be achieved in one or two courses. Or consider the opposite: a business major adds a computer science minor, in order to be able to use the computer more expertly in business applications. And here is another alternative: this same person decides to major in both business and computer science, thereby achieving highly developed capacities in both disciplines. The career opportunities and the alternatives that are open to this individual are very wide and promising. This student can design business applications systems that represent a maximum communion (or dialogue) between business and computer science. For in applications work, it is necessary that the programming specialist have a very good understanding of the other discipline; ideally the applications programmer should be knowledgeable in the second (applications) area.

We see, in the foregoing examples, applications that require only a few hours of training, and areas that require more preparation, ranging from taking one course, to several courses, to a minor in a second discipline, and finally to becoming proficient in two areas.

Examples of Applications

One of the most detailed surveys of computer applications in fairly brief form is the thirteen-page summary found in the "Computer Directory and Buyers' Guide" (4). A listing of applications is divided into four main groups:

    I. Business and Manufacturing in General
   II. Business—Specific Fields
  III. Science and Engineering
  IV. Humanities

Each main group is subdivided into an alphabetical listing of areas or fields, and each area is then reviewed with an alphabetical listing of applications. In area II, for example, there are 25 fields, with an alphabetical listing of applications in each. In the four main categories, 2672 applications are listed. Table 6.1 is a breakdown of the four main categories, together with subcategories, and individual figures for each area, along with the totals.

6.5 THE SOCIAL IMPLICATIONS OF COMPUTERS

The uses of computers affect people's private and working lives, their leisure patterns—and even affect man's image of himself. When we discuss the "social implications of computers" we are referring to the ways in which people's lives are transformed. Here are some important topics:

   —Computer Literacy
   —Changing Work Patterns, Automation, Obsolescence
   —Changing Educational Patterns—New Jobs, New Places
   —The Leisure Ethic vs. the Work Ethic
   —Privacy and the Individual
   —The Technological Imperative
   —The Social Responsibility of Every Person
   —The Social Responsibilities of Professional People
   —Computers and Man's Future

Although the computer touches our lives in innumerable ways, the

TABLE 6.1

NUMBER OF COMPUTER APPLICATIONS PER FIELD

Identi-fication	Field of application	No. of computer applications listed here	Identi-fication	Field of application	No. of computer applications listed here
I. *Business and manufacturing in general*			(*Science and engineering. continued*)		
1.	Office	121	4.	Botany	3
2.	Plant and production	79	5.	Chemical engineering and chemistry	36
	Subtotal	200	6.	Civil engineering	89
			7.	Ecology	10
II. *Business—specific fields*			8.	Economics	8
1.	Advertising	22	9.	Electrical engineering	36
2.	Automotive industry	54	10.	Geology	14
3.	Banking	46	11.	Geophysics	7
4.	Educational and institutional	101	12.	Hydraulic engineering	27
5.	Farming	44	13.	Marine engineering	31
6.	Finance	41	14.	Mathematics	44
7.	Government (local, state federal)	210	15.	Mechanical engineering	37
8.	Health and medical facilities	89	16.	Medicine and physiology	191
9.	Insurance	44	17.	Metallurgy	3
10.	Labor unions	11	18.	Meteorology	25
11.	Law	31	19.	Military engineering	38
12.	Libraries	12	20.	Naval engineering	7
13.	Magazine and periodical publishing	26	21.	Nuclear engineering	19
14.	Military	12	22.	Oceanography	23
15.	Oil industry	83	23.	Photography	15
16.	Police	29	24.	Physics	24
17.	Public utilities	58	25.	Psychology	27
18.	Publishing	11	26.	Sociology	5
19.	Religious organizations	62	27.	Statistics	27
20.	Sports and entertainment	86		Subtotal	915
21.	Steel industry	31	**IV.** *Humanities*		
22.	Telephone industry	21	1.	Anthropology	2
23.	Textile industry	16	2.	Archeology	17
24.	Transportation	174	3.	Art	11
25.	Miscellaneous	157	4.	Games of skill	9
	Subtotal	1451	5.	Genealogy	3
			6.	Geography	4
III. *Science and engineering*			7.	History	13
1.	Aeronautics and space engineering	126	8.	Language	14
			9.	Literature	10
2.	Astronomy	16	10.	Music	19
3.	Biology	30		Subtotal	106
				TOTAL	2672

SOURCE: "Computer Directory and Buyers' Guide" (4).

majority of the people living in the United States do not understand what it can and cannot do. Because people have not used a computer, they are not able to form reasonable reactions to this complex machine.

We are in a transitional period, in which computers are drastically changing our lives, and we need to know personally about these machines in order to make constructive uses of the power of the computer, and in order to legislate sound practices and laws regarding its uses.

Computer Literacy

There is one remedy for dispelling the fear and prejudices about computers, and that is computer literacy. This means an introduction and beginning familiarity with computers and how they work. It means successfully using a computer to perform some work, and to solve problems.

Due to fear and lack of positive exposure, a great many people have a strong, negative attitude towards computers and technology. Some think that the computer is dehumanizing, and that activities that do not make use of the computer are more human. This is, of course, untrue.

The computer will do what it is told to do, provided the instructions are given in such a way that the computer is able to accept and process the instructions provided by a person. It is neither good nor evil. What is "good or evil," if we may use these terms, is the programming, the intent, and the actions of persons using the computer.

In this brief chapter, you have experienced a "mini-exposure" to computers. Perhaps as a result of reading this chapter, you will take an exposure course in computers, and begin to find ways in which you can begin your "acquaintance" with computers. The references cited in this chapter will offer suggestions for this introduction to computers.

At the present time, all universities and colleges offer many types of introductory courses about computers that can be stimulating and useful (14).

High schools are beginning to introduce computer languages, and in the next five years, "Introduction to Computers" and programming languages will be common. Since there are far more alternatives for higher education than before (trade schools, institutes, junior colleges, state colleges and universities), it is desirable that all people have computer experiences in order to prepare them for their adult lives, for their career goals, and for socially enlightened behavior in a technological society (10).

There is at present a great need for adult education courses dealing with computers and man. So far, the junior colleges in this country have been the leaders in offering tuition-free courses of this type.

There is a growing recognition in higher education that all people need to study and know more about technology and how it affects mankind and the earth. There is a "grassroots" movement in today's colleges and universities to consider and teach courses about the social implications and the effects of technology. According to Ezra D. Heitowit, editor of a recent publication titled "A Listing of Courses and Programs in the Field of Ethical and Human Value Implications of Science and Technology," published by the Program on Science, Technology, and Society of Cornell University, Ithaca, New York, 384 institutions reported activities in Science, Technology and Society areas; 192 institutions reported some teaching activity related to ethical and human value considerations of technology. The authors of this chapter have worked ardently in this area of computers, social responsibility, and man (5, 9). References in these very recent publications should be useful in further study of these significant concerns.

But perhaps you have more immediate concerns, about people's jobs, their privacy, or leisure, etc.

Changes in Present Occupations and Education

In section 6.4, brief mention was made of what we called "user-oriented" computer systems, and how they changed (and improved) some jobs. For a brief period of time, people said (and some still say): "Computers will only make more jobs. They are not going to put people out of work." This is not totally true.

Although computer applications are becoming so common as to be found in literally almost every human endeavor, often these applications are not extensive. However, in the very near future these applications will become more extensive and pervasive. An important idea is that as computers become more commonly used by more people, these applications will be an aid and not a hindrance to people in their working and private lives. Another important idea is that computers will increasingly affect and alter many occupations. Here are a few examples.

Training for Varied Occupations. Education, or practical job-oriented training, will become increasingly computer-aided. Even at the present time, individuals with special computer applications skills are receiving more favorable positions than those without these skills. A few examples are: statisticians with experience in automated statistical packages and in computer-aided graphing; geographers with automated cartography and statistical experiences; engineers and draftsmen with computer-aided design capacities; teachers with backgrounds in computer-aided instruction.

With declining enrollments in colleges and universities, and with fewer college graduates obtaining jobs after completion of their studies, there is more emphasis on *getting a job after graduation*. Government and industry are specifying far more what they are looking for in graduates, and higher education is quickly adopting these suggestions. Jobs of today and tomorrow require computer applications capacities in almost every field.

In order to prepare themselves for an open, growing future in the working world, students need to acquire job-related computer skills before graduation.

A discussion follows, reviewing changes in many kinds of jobs.

Unskilled Workers. This term signifies persons holding jobs that require a relatively short period of training and job skills. There are many levels of skills in the category of "unskilled workers," and the work that is accomplished is important. However, at the present time, and in the very near future, people who do not have medium to high levels of jobs skills will find that there will be far fewer jobs available to them. New factories are able to utilize automated production systems more readily than established factory systems, because of the cost that would be required for the latter to change over to the new systems.

In time, many of the manual, less skilled jobs will become obsolete. However, there are many unskilled jobs that will remain as they are, because the cost of automating these positions would exceed the cost of people performing these same tasks. Driving a school bus is one example. Road work is another.

It is important to remember that man has been performing many tasks that were unpalatable, uninteresting, and even less than human. Many people have had to take certain kinds of jobs just to stay alive, to survive—and these jobs were not ones one would ideally choose. With increasing automation, many of these "unwanted" drudgery jobs will be eliminated.

Clerical Workers. In the applications section, we reviewed levels of computer uses in banks and offices. With the decreasing cost of computer systems, and the availability of user-oriented languages, clerical positions will soon be upgraded and altered. People who do not have this type of training will have fewer job opportunities. However, as education at *all levels* becomes more up-to-date, people will be prepared for these computer-aided clerical positions.

Here are some clerical jobs that will be greatly changed within the next ten years: scheduling, inventory, accounting, bookkeeping, correspondence production, text editing, graphing, report writing, and mail-out production work.

The processing of numerical data in offices and business is increasing exponentially every year, and the processing of clerical data, which includes both word and numerical processing, is following a close second.

All large and medium-sized firms have word-processing centers at the present time. Although on-the-job training is altering the general clerical positions, soon high schools, junior colleges, and colleges will offer specific courses in computer-aided clerical systems for varied levels of office work. The old, manual methods are on the wane, and will soon be obsolete.

The Professions. People have different ideas of what constitutes a "professional." One criterion is advanced academic degrees, to at least the master's level, and generally on to the doctorate level.

> What separates a professional from a non-professional (or a para-professional) is that the professional is responsible for constructive use of advanced technical (or specialized) capacities, and that in this function, the professional directs the work or process. The professional is assisted in this performance by important but subordinate aides. The professional makes the final decisions, and subordinates carry them out. (10)

Most people would consider at least the following persons as professionals: engineers, doctors, lawyers, teachers, nurses, architects.

The professional person will be freed from some of the time-consuming tasks of the present. The use of terminals and minicomputer systems will make present applications even more common and extensive in the professional world. Here are a few examples: case histories, medical diagnosis, monitoring of patients, and computer-assisted operations in hospitals; search for precedents, text editing, case scheduling, and computer simulation of cases in law; computer-assisted and computer-managed instruction in all levels of education; computer-aided design, testing and simulation of buildings, bridges, etc. in architecture and engineering.

Paraprofessionals will be trained to accomplish specific levels of tasks to aid the professional, and these persons will make greater use of computers in their work.

Here are some benefits of the "enhanced" professions. With the aid of computers and paraprofessionals, doctors will be able to help more people. Preventive medicine will become feasible and economical, just like maintaining a car before it breaks down. The scheduling of cases in law courts will remove the time lag between the time of an accident and settlement of a case. The poor will not lie languishing in jails, awaiting trials. People will be able to ascertain, via the computer, the status and prospects of a case, and come to agreements and settlements more quickly. With the availability of quick polling of constituents on the local, state, and federal levels, revisions in the law will be facilitated.

In general, people in the professions will find their positions upgraded, and they will be able to use their skills to help counsel and serve more people than they can at the present time.

One of the greatest changes will be in teaching, where varied media presentations will allow students to study the curriculum at their own pace, to test themselves, and to proceed at their own comfortable rate, rather than at the present pace, which is geared for the middle learner, not the fast, nor the slow. Mastery-learning-oriented systems will teach people specific skills, upon which they can build sequentially. Almost all colleges and universities are offering experimental courses in Learning Resource Centers. Cable television and educational networks are offering accredited TV courses, and there is a firm establishment of a trend towards video-assisted instruction and computer-assisted instruction at flexible time periods for many kinds of people (3).

Education is the one professional field that is lagging behind in making use of instructional-technology systems. There are reasons: computer systems are still too costly for many schools; computer-assisted learning software systems are often not transportable (to be used in any location); and learning systems are not sufficiently oriented toward general users at the present time. All these things are changing right now, and when other changes occur, we may anticipate that teachers will overcome their present fear of being replaced by becoming more adventurous and freeing themselves of redundant tasks and paperwork. There are enough solid, established computer-aided learning centers to be able to make predictions of changes, without exaggerated promises for education. Eventually, teachers will have far more time to communicate with students on a personal, individual basis—because of computer applications.

Job Changes in a Lifetime. What could happen to most people in an adult working lifetime is that their skills become upgraded and they experience more stimulating working lives. What is inevitable is a reduction of the work week, with a salary sufficient to live comfortably, but with fewer hours of work to be performed, in order that more may work (6).

Young people in high school and college at the present time will have several job changes in their adult working lives, with perhaps four or five changes in positions (16). We can see this trend now, where many forty-year-old people begin a second career, and many sixty-five-year-olds begin a third career. This trend is also visible in the lives of many women who work before marriage, then remain at home until their children are in school, returning to school and working, then changing jobs sequentially, upgrading their skills.

The idea that education will prepare one for one lifetime job is obsolete even now, and with increased computer applications, job changes for everyone are inevitable. What is required is flexibility, and life-long learning. Paul Armer has written of "professional obsolescence" in which scientists and engineers find that the flood of new information in their fields makes it difficult to maintain a professional career and to keep up with new ideas (1).

What does this mean, then? A broader education is required, and more training in general problem-solving is necessary for everyone. People need to be educated to learn more material faster, to retain it, and to continue life-long learning. This is where the role of the colleges and universities will be changed; they will increasingly be oriented towards learning for new careers, and learning for leisure as well. Thus education may simultaneously become more practical and more humanistic. This trend is visible in all levels of education where people are taught to enjoy life more: recreational crafts, film appreciation, photography, dance therapy, flower arrangement, and painting for pleasure.

Job sabbaticals at many levels are becoming more common, extending beyond the common teacher sabbaticals, where people are paid to learn new skills and broaden their perspectives (8, p. 190).

Toffler repeatedly refers to the alternatives open to people in the future and the necessity for cultivation of an open, flexible attitude (17).

Employment Changes. With the flexible patterns now beyond the fad phase, it is becoming common to choose hours of work. With the lowered work-hour requirements that are inevitable, people will have time for leisure and to prepare themselves for potential new careers. Retraining by government and industry will become more common. We can see this trend in various agencies, such as Departments of Rehabilitation, and in industry's own training programs to help people learn new techniques to perform new tasks (8, pp. 185–88). An example of employment changes is the highly organized training programs of state and federal civil service, in which people are constantly given new training for upgraded positions. Many firms have paid employees' tuition to attend special schools and conferences and to work on advanced degrees useful in their work. The trend towards organizational sponsorship of education for job changes is firmly established, and computer-applications education will become even more important in the next ten to fifteen years.

Databanks for Employment. Another area of computer applications will be developed to help people change jobs and to prepare themselves for changes. One of the computer's beneficial services will be in the form of national and regional databanks for employment, so people can move smoothly from one job to another. There will be a greater trend toward mobility (17) and individuals may look forward to changes in jobs and locations, experiencing other climates and areas rather than remaining in one place, one job for a working lifetime.

Services for People. With the necessity of manufacturing items that last longer in order to conserve our resources, there will be an emphasis on repair and maintenance of equipment, and the need for specialized services for people. Many of these new jobs will be computer-aided. Serving people will become more important and profitable than creating products for obsolescence (11).

Life-long Learning. Many university and college extension departments use this meaningful title: Learning for a Lifetime. It indicates that learning should be continuous, for every human purpose and desire, and that learning is not the prerogative of the young or of the financially or intellectually endowed. All people will have the opportunity and the time to be creative, philosophical, and practical. In professional curricula at the present time, higher education places an overemphasis on technical courses; in the humanities, there is an overemphasis on creativity and self-expression, and a neglect of the important idea of how to earn a living (in addition to being creative and philosophical).

With the aid of computers and other technological systems, a great revolution will be strongly visible in the next ten years, and the opportunities open to all kinds of people for intellectual and practical learning will be one of the main gifts of computers and technology.

The Leisure Ethic vs. the Work Ethic

Along with reduced working hours and the manufacture of long-lasting items, we will see the appearance of a new ethic called the "leisure ethic" that will replace the present emphasis on work. In the United States, a competitive, goal-oriented culture has placed great importance on working assiduously, with the result that even leisure is pursued "assiduously," often resembling a form of work. Definitions of leisure explore the differences between spare time used for one's work and/or status, and true recreation or leisure—"the freedom from the necessity of being occupied." The refreshment of the mind, the body, and the soul are the goals of leisure. Leisure is not "watching television, shopping, fixing up the house, keeping fit, or taking adult education classes" (8, pp. 257–62).

There have been many interesting studies in uses of leisure—added free time—and they indicate that people are *not prepared for leisure,* and that there is a need to re-educate people *away* from the necessity of being occupied at all times, of doing useful things continuously, and to grow away from this literal "programming" of our culture.

It is important to recognize that people are very different in their leisure-time pursuits. A scientist may grow vegetables as a form of leisure; a blue-collar worker may study Great Books; a high-school teacher may play a Renaissance instrument; and a surgeon may enjoy ceramics in his free time. Leisure is any activity (or inactivity) that refreshes the person, that re-creates the individual. Recreation might even be computing for pleasure!

The enjoyable use of leisure is an important social issue arising from increasing uses of computers in myriad applications. If people can be assisted in enjoying free time, then the added leisure afforded by technology can be a great boon for mankind. We cannot evade the necessity of reduced working hours for all human beings, because technological systems will perform more tasks for man, and perform them quickly and inexpensively.

Taviss has predicted that in the year 2000 people will work fifteen hours per week (16), but she is perhaps excessive in this prediction. Perhaps twenty-five hours instead of the present forty work hours would be a more prudent prediction.

In summary, preparation is needed for this leisure phase of man's new life. This is difficult for many people, as evidenced by the high death rates among retired persons soon after retirement, for they have lost their "reason for existence," *work*. The courses mentioned earlier, in which man, technology, and human values are reaffirmed, can make a difference in preparing people for more bountiful lives.

Even now, young people want meaningful jobs, not just survival as in generations past. They want more freedom, more stimulation, and less drudgery.

Humanely designed computer systems can provide this new freedom for man. People will have time to build, to create, to think, and to be individuals.

Privacy and the Individual

Before the invention of computers, the gathering of data and information was slower and more laborious, and the dissemination of information was less efficient. A person was more "private"—less was known about a person than at the present time. With the establishment of various forms of data banks, much more information is quickly gathered, sorted, retrieved, and disseminated about people to other people. In addition, much unncessary information is gathered in the process, simply because the computer affords the capacity to process, store, and output information. We have not addressed this vital issue of gathering only "necessary" information. We are often computing to compute.

Here are a few uses and users of data systems: hospital records; elementary, high school, college, and university record-keeping; social security; banks and firms that disseminate credit cards for innumerable purposes; federal and state agencies, including income tax and retirement; department stores and mail-order houses that give credit to customers. With the growing ease of establishing data systems and disseminating information, we may eventually move towards a "womb-to-tomb" dossier society.

Many countries have established privacy legislation and enforcement agencies. Many states are enacting specific privacy protection regulations. Special boards are studying the privacy rights of individuals, and are seeking to establish a common-sense solution that allows for automation of information, but that also protects specified rights of the individual. The reader is urged to review the dictionary definitions of "privacy" and related synonyms, to realize the implications of civil rights, and the "privy of person" that is being changed at the present time. A current definition of privacy is: "Privacy is the right to control the dissemination of information about one's self" (12, p. 25).

The pros and cons of a national data center are well explored by Arthur Miller in his text *Assault on Privacy* (12, pp. 54–67). Another extensive study of computers and data banks is the superior work by Alan Westin, *Databanks in a Free Society—Computers, Record-Keeping and Privacy* (18), sponsored by the Computer Science and Engineering Board, National Academy of Sciences. One of the revealing insights of this text is that organizations that did not respect the rights of individuals as persons in their manual record-keeping dossiers tended to invade the privacy rights of people, and those organizations that were concerned with the civil rights of human beings tended to establish automated systems that did not invade the rights of their clients.

The fact that computer applications of gathering, processing, and disseminating information are developed sufficiently so that there is a proliferation of regional and national data bank systems, implies that we may have reached the "point of no return" regarding data banks. This is one important reason why computer literacy of every person is needed, to control and legislate constructive uses of data systems that benefit people and give people some degree of privacy.

The old definitions of privacy, leisure, and work are drastically altered. Legislation regarding computers and their applications, regarding dissemination of information, is a great social issue that will not wait until we all have computers in our homes. In addition to being enlightened about computers, this means participation by people in their government, and there are many texts on this topic, including two well-known works by Toffler (17) and Meadows (11).

We cannot rely upon the computer people to protect us in this area of privacy. Although professional organizations are urging their members to "socially responsible" codes of behavior, professional preparation for computer scientists and engineers is primarily technical, with insufficient emphasis on professionalism and social responsibility, which is our next topic.

Social Responsibility and Professionalism

> Social responsibility is the knowledge that one is responsible for one's actions, and the willingness to assume this responsibility. A mature person is "socially responsible." A professional person is socially responsible. (10)

There have been many discussions of technology and social responsibility. One of the most moving and revealing sources on this topic is the book *Einstein on Peace* (13). Here the important questions of the responsibility of *every person* and the responsibilities of scientists are discussed at length, brilliantly.

At times, organizations have attempted to establish a code of ethics for their members. If the member violates this code of ethics, the person is barred from membership, and from practice of the given profession. We are most familiar with the codes of ethics of doctors and lawyers. But man has been attempting to define a code of ethics or behavior for thousands of years, and records date back to 2000 B.C., a total of four thousand years.

Where does the individual's social responsibility begin and end? Where does a professional's social responsibility begin and end? There is a growing concern in the scientific-professional community for establishment and definition of codes of social responsibility. The magazine *Computers and People,* the oldest computer magazine, has for twenty-six years been most articulate on this vital issue of social responsibility (5). Dr. Charles Susskind has proposed an "Engineer's Hippocratic Oath": that an engineer use his or her scientific knowledge for the benefit of mankind and the earth (15). Readers are urged to study this Hippocratic Oath, and to formulate one for themselves.

The Technological Imperative

Science has assumed an attitude of total freedom in the exploration of new discoveries. "Scientific truth" has been considered a form of pure knowledge, a form of "good." "The technological imperative is the force to implement a new technology because it is possible to do so. Man cannot, in the case of an imperative, decide not to implement the new technology, because it must and will be done" (7). Earlier, we referred to the point of no return. But have we reached that point? Is it possible to call a halt to some forms and some applications of technology? Are we caught up in a whirlwind of technology that has swept us away, and have we become mute, impotent fragments in this maelstrom?

The grassroots movement of people considering what technology is, what it should be, and where it will go, is one important denial that we are caught up in a "technological imperative." Because of the innumerable important scientific discoveries within the last two decades, there is renewed interest in and emphasis on thinking of implications of applications before implementation. True, this is only a beginning, but at last there is perceptible a new era, in which people are beginning to think and to act about the consequences of their actions, their science, and their intentions. Because of the complexity of present-day life, mankind is being forced to regard these great issues, and to assume the mature burden of being socially responsible.

Must we use technology and computers for potential purposes, merely because they are there, much as the mountains are climbed, because they are there? Can we continue to wage war with computers and to control bombs and missiles with computers, and to use sensors for surveillance, controlled by computers? Who is responsible for these atrocities? Who will accept the burden?

It is important to remember that man has been waging war for thousands and thousands of years. In the past, man accepted this "grim necessity." But with the invention of the atom bomb, computers, and other forms of technological discoveries, war has become so powerful that at last man has been forced to survey this bestial part of his heritage, and to begin to perceive that it does not solve problems—and that perhaps we need to think of a humane way to solve the problems of mankind. It is paradoxical that with the present technological complexity of warfare and its potent capacities for annihilation, we are forced to think of *not waging war,* because we cannot afford to take the chance of our death.

The constructive use of computers and technology is the most important question that must be faced by mankind. Every person in a society is responsible for constructive (and destructive) uses of technology. We are not an island unto ourselves anymore, but a global village, and what affects one, affects all. The authors of this chapter have chosen not to evade this very important question of constructive uses of computers and technology. And the reader cannot evade this question. To become knowledgeable is a partial answer. Another is to legislate honorably and intelligently on important issues. Still a more complex answer is to know who you are, where you are, and to act responsibly in every action, every day. For only people can control the technological imperative. Only people can devise and approve of uses of technology. We cannot retreat into the past and negate the potential benefits of computers. Each person will have to stand to be counted. This is the essence of social responsibility.

Computers and the Future

In the introduction to *Limits to Growth,* U Thant states that we can see the shape of the future in the present—we can see the direction of the world in the present, and within the next ten years, we will see the shape of its future (11).

In present-day applications of computers, which have permeated every discipline, almost every area of life, we can see the shape of computers and the future. We can predict reliably that hardware will be very inexpensive, that people will use computers in their cars and their homes, and that the applications will multiply exponentially. The size of computers will become smaller and smaller. The cost will be lower and lower. Languages for computers will be user-oriented—and computers will mature to the point where they hopefully will understand our imprecise human languages. Computer memories will become larger and more extensive. "Prediction . . . every defined intellectual operation will be done faster and better and more reliably by a computer than by a human being" (2). The key word in the last sentence is "defined." More tasks and problems and activities are being defined by man since he invented computers. We know more about what we were doing before, since we were required to define it, in order to use computers. This new knowledge is tremendously important. And this new capacity to define ideas is leading to new knowledge, and new insights.

For at least a decade, there has been much discussion about the gap between the humanities, people, and science. This is even more true now. And the area of required work is to develop people and human communications, so as to lessen that gap.

The futurists have differing views of what the world holds for mankind. A good proportion of these prophets of the future have dire predictions for man's survival. Many people use the argument that man has always managed to survive, somehow. This is not a sound basis for being optimistic about the future.

We have reason to be optimistic about computers and the future for many differing reasons:

—More people are learning to use computers, and the fear and mystery are lessened.
—More people are learning about technology, social values, and man, and with this new attitude and knowledge, they may act more humanely.

—We have more complex tools with which to solve the problems of mankind and the earth.

—The computer has served as an extension of man's mind, and with this added capacity, man has the potential to act more rationally, to solve more problems than ever before.

—We are aware of the importance of people and the earth, and are becoming aware of the necessity for social responsibility.

—We are learning to think.

While this may not be the "best of all possible worlds" as Candide admonished, this is the only world we know. What use we make of computers and other forms of technology in this only world depends upon every human being who lives in this only world.

Computers and the future are our responsibility. The future of computers, and the shape of our future, is our invitation, and our gauntlet.

6.6 EXERCISES

1. Write a short comparative essay on the first and second (computer) Industrial Revolutions. Which brought more radical changes in human life?

2. Search out and name a dozen scientists and engineers (in chronological order) who most influenced computer technology.

3. Discuss whether or not a computer can think.

4. Name a half-dozen programming languages and discuss their main characteristics.

5. Write a short essay on a particular type of computer that you are most familiar with.

6. Do you foresee the possibility of "computer utilities" similar to gas and electric utilities in the near future? Discuss.

7. Find out how much electric energy a medium-sized computer consumes per hour and compare it with an ordinary electric range for home use.

8. Discuss the advantages and disadvantages of computer-assisted instruction.

9. Do you agree that computers cause unemployment? Discuss with facts.

10. Write a short essay on how computers may affect the life of housewives.

11. Do you agree that computers will allow more leisure time for mankind? Discuss briefly.

12. How do computers interfere with our private life? Write a short essay discussing your views.

13. Do you foresee that computers will eventually understand human language? With a library search, write a short essay supporting your opinion.

14. Do you think that man will surrender to computers and be manipulated by them? Can you cite some examples?

6.7 BIBLIOGRAPHY

1. Armer, Paul. "The Individual: His Privacy, Self-Image and Obsolescence." Proceedings, Meeting Panel, Science & Technology, 11th "Science and Astronautics," p. 129. Washington, D.C.: U.S. House of Representatives, U.S. Government Printing Office, 1970.
2. Berkeley, Edmund C. "Philosophy and Computers: Some Personal Views." *Computers and People.* vol. 26, no. 5, May 1977, pp. 16–18.
3. Bork, Alfred. "The Physics Computer Development Project." *Computer Graphics and Art.* vol. 1, no. 3, August 1976, pp. 10–16.
4. "Computer Directory and Buyers' Guide." Berkeley Enterprises, Inc., Newtonville, MA 02160. Published yearly.

5. *Computers and People* (formerly *Computers and Automation*) vols. 1–26. Edmund C. Berkeley, editor. Published by Berkeley Enterprises, Inc., Newtonville, Mass.

6. De Grazia, S. *Of Time. Work. and Leisure.* New York: Twentieth Century Fund, 1962.

7. Dorf, Richard C. *Technology. Society and Man.* San Francisco: Boyd & Fraser, 1974; p. 107.

8. Gotlieb, C. C., and A. Borodin. *Social Issues in Computing.* p. 190.

9. Hertlein, Grace C. "Computers, Society, and Man." In *Proceedings.* Computers in the Undergraduate Curriculum, Washington State University, Pullman, Washington, June 1974.

10. Hertlein, Grace C. "Engineers, Computers, Social Responsibility." In *Proceedings.* COMPCON, IEEE Computer Society, Williamsburg, Va., June 1977.

11. Meadows, D. H., et al. *The Limits to Growth.* New York: Universe, 1972.

12. Miller, Arthur. *The Assault on Privacy.* Ann Arbor: University of Michigan Press, 1971; p. 25.

13. Nathan, Otto, and Heinz Norden, eds. *Einstein on Peace.* New York: Simon & Schuster, 1960.

14. *Proceedings.* Computers in the Undergraduate Curriculum. Dr. Ted Sjoerdsma, Lindquist Center for Measurement, University of Iowa, Iowa City.

15. Susskind, Charles. *Understanding Technology.* Baltimore: Johns Hopkins Press, 1973.

16. Taviss, Irene, ed. *The Computer Impact.* Englewood Cliffs, N.J.: Prentice-Hall, 1970.

17. Toffler, Alvin. *Future Shock.* New York: Random House, 1970.

18. Westin, Alan F., and Michael A. Baker. *Databanks in a Free Society—Computers. Record-Keeping and Privacy.* New York: Quadrangle, 1972.

Chapter Seven

POLLUTION

7.1 WHAT IS POLLUTION?

Numerous scientists, commissions, congressional committees, professional organizations, and others have wrestled with the problem of defining pollution. Some of the more imaginative or most widely used attempts are discussed below.

In an ecosystem, life is supported by interrelationships among the organisms. The interrelationships are called a "food web," which indicates which organisms are the food source for which other organisms. The more diverse the types of organisms in the ecosystem and the more complex the food web, the more stable the ecosystem. For this reason, ecologists like to speak of pollution as "anything that decreases diversity" and thus creates a less stable ecological system. Problems arise, however, when some pollutants (e.g., heat) can actually increase diversity.

A more practical definition might be "the presence of anything in undesirably large concentrations," recognizing that some polluting substances are harmless or even beneficial in small quantities (12). The problem in this case is with the word "undesirable." A concerned public and General Motors view the number of cars on the road in distinctly opposite ways, for instance.

Legal definitions of pollution are usually based on some arbitrary standards, such as "the dissolved oxygen level shall not be less than 4 mg/l." *

---

* mg/l (milligrams per liter) represents the stated milligrams of a chemical per liter of water.

While 4.1 mg/l is acceptable, 3.9 mg/l is not. The problem is that not only could a fish not tell much difference in 0.2 mg/l of dissolved oxygen, but our measuring methods are usually less precise than 0.2 mg/l.

One widely quoted definition of pollution, often used by engineers to justify exploitation of resources, is "the use of a resource so as to interfere significantly with another beneficial use." * By this definition, if the use of a stream as an open sewer is its most *beneficial* use, then the stream is not polluted.

This latter definition assumes that the world's resources are here for people to use, and does not acknowledge the existence of our coinhabitors, or the preservation and conservation of natural habitat. It is an action-oriented definition, asking only for a definition of the key words "benefit" (dollars—what else?) and "significant." The latter could also be expressed in monetary terms. For example, the benefits of white-water canoeing are not very significant when compared to the benefits of a hydroelectric project. Most people, however, would have strong reservations about such value judgments.

One congressional staff committee defined pollution as "anything that degrades the environment," neatly defining one difficult word by using an even more obtuse word (environment) in the definition.

Another government commission based its entire report on the premise that pollution is "anything discharged into water, air or land." A cow in a meadow would thus qualify, much to its dismay.

Both of these definitions of pollution represent the "clean as possible" or the "zero discharge" approach. By defining pollution as anything discharged into the environment, the problem of definition is neatly solved, but the use of this definition in a practical sense is difficult. We would certainly agree that a glass beer bottle thrown on the grass in a public park is pollution (litter—a visual and potential health problem) but this same bottle properly disposed of is also pollution by the above definition.

One author has defined pollution as "the wrong thing in the wrong place at the wrong time" (15). This overly simple definition suffers from the fact that it doesn't take into account naturally occurring conditions. For example, pollen is certainly the wrong thing for a hay-fever sufferer, but one would not set out to eliminate all the pollen in the air just because some people sneeze.

---

*Private communication from J. C. Lamb III, Professor of Sanitary Engineering, University of North Carolina, Chapel Hill.

In most cases, no attempt is even made to define the term. For example, a recent annual report (7) of the Council on Environmental Quality introduces pollution with this paragraph:

> Pollution in its various forms—air, water, noise, solid waste, and hazardous substances—has been an environmental concern for many years. In general we all agree that pollution should be reduced. Differences arise over how serious a problem one particular type of pollution may be, whether the technology to abate it is available, and whether the degree of abatement achieved is worth the cost.

The word itself is never defined, and it might not be at all unreasonable to ask exactly what it is that we are abating at such cost.

The Oxford dictionary offers that to pollute (v.) is to "destroy the purity or outrage the sanctity of." There is much to recommend the concept that pollution is in fact the despoilment of a sacred trust. The earth which we have inherited and will so quickly bequeath to our children—the very finite spaceship—is indeed ours to destroy or protect.

We will come back later to this question of the sanctity of nature. Before us now, however, is the need to establish a functional definition of the title of this chapter.

We would like to argue that pollution can be defined in a broad context as the introduction into the environment of a waste, a material or energy having no further value, as a result of human activity so as to cause a stress or unfavorable alteration of an individual organism, a community, or an ecosystem (17). Using this definition, we can readily classify chlorinated-hydrocarbon pesticides as pollutants (even though they might also be beneficial). Similarly, fluorocarbons used in spray cans, which in all probability adversely affect the ozone layer protecting the earth from ultraviolet radiation, would be a pollutant. On the other hand, a cow doing its business in a meadow, or effluent from a properly operating septic tank, would not be classified as pollution.

No definition is perfect, and one could certainly puncture holes in this one. It nevertheless will serve here to emphasize that pollution

—is caused by people;
—is a waste (a residual);
—can be a material or a form of energy;
—represents a loss of value (to health, ecosystem, etc.)

We will thus speak of such pollution for the remainder of this chapter, returning to the question at the conclusion.

7.2 WHAT CAUSES POLLUTION?

In a natural system, nothing can be a waste—a product with no value. If such a waste had ever existed, the world would soon have been covered with it and this waste would have choked off all forms of life. Natural systems run in cycles, where the waste in one system is used as a raw material by another. For example, plants produce oxygen as a waste which is used by animals, which in turn produce carbon dioxide as a waste, which is used by the plants.

The combination and interweaving of many such cycles into one dynamically balanced process is the secret of an ecosystem.* The term "ecology," coined by Ernst Haeckel in 1869, combines two Greek words which together can be loosely interpreted as "the study of a place to live." Since we all live in an ecosystem, ecology is thus·the study of the form and function of ecosystems.

Ecosystems can be quite limited in scope (e.g.,. a snail, plants, and bacteria in a bottled terrarium) or unbelievably complex. In a strict sense, an ecosystem does not have externalities—*everything* influences it in some small but nevertheless finite way. Although the snail-plant-bacteria balance appears to be self-contained, in truth the plant requires outside light, the jar has to transfer heat, etc. One could thus rightfully consider the entire earth as an ecosystem—a perfectly balanced incredibly complex series of cycles which allow the continuation of life.

It is possible, however, to disturb an ecosystem sufficiently to upset the balance and destroy the system. For example, putting a terrarium in a dark room will kill the plants, eliminate the supply of oxygen for the snail, and eventually stop all life.

Fortunately, the global ecosystem is not so fragile. This system is so well-balanced, and is able to absorb such large shocks with only minor readjustments, that there should be little concern that the system can be destroyed.

But this fact was true only before the emergence of the most destructive creatures ever known to have existed on the face of the earth—organized people. For example, it has been stated that human beings expend in bodily energy about a thousand kilowatthours (kwh) per year (3). Prehistoric man, therefore, utilized energy at about that rate. Currently in the United States, energy is consumed at a rate of ninety thou-

---

*For further readings on ecosystems, see for example A. Turk et al., *Environmental Science* (23), or for a more rigorous treatment, E. P. Odum, *Fundamentals of Ecology* (18).

sand kilowatthours per person per year (6). The impact of technology on the world ecosystem is clear.

It bears repeating that before people organized themselves into communities where division of labor, quest of material wealth, and ownership of resources became accepted modes of behavior, there *was no pollution!*

In Pogo's words, "We have met the enemy, and he is us."

Our life on this planet has upset the ecosystem, and the allowable degree of perturbation before the damage is permanent (and fatal) is open to argument. But how did we do this? What is it about our manner of inhabiting the planet that has produced such potential disaster?

Although other answers to this question can perhaps be advanced, we would like to argue that our inhabitation has been destructive for four reasons:

1. There are too many of us.
2. We want to live in compact communities.*
3. We have a system where consumption and wasteful use of natural resources is a driving force of civilization.
4. We exploit our technological skills.

Let us consider these individually as the four primary answers to what causes pollution.

Population Growth

A combination of our cleverness and our ability to exploit resources has resulted in a human population growth of incredible proportions. It is estimated that human population remained essentially constant from ten thousand years ago to perhaps five hundred years ago, at which time the scientific-medical-industrial revolutions abruptly changed the curve. The increase from perhaps five million people five hundred years ago, to about four billion today all occurred within but a moment in geological time. Table 7.1 shows the growth rate of the population, estimated for various points throughout history. Current estimates indicate that the human population of the earth will double in thirty-five years. Should this trend continue (presently adding the equivalent of a new city of two hundred thousand people *each day*), the earth will simply not be able to support life.

---

*The reader is advised to refer to chapter 2 for new trends in habitation.

These four billion people all produce waste. Worse, they produce it in concentrated (urban) areas. The amount of this waste (domestic, industrial, agricultural, etc.) is simply overwhelming the environment.

TABLE 7.1

DOUBLING TIMES FOR WORLD POPULATION

Date	Estimated world population	Time for population to double
8000 B.C.	5 million	
		1,500 years
1650 A.D.	500 million	
		200 years
1850 A.D.	1,000 million (1 billion)	
		80 years
1930 A.D.	2,000 million (2 billion)	
		45 years
1975 A.D.	4,000 million (4 billion) Computed doubling time around 1972	35 years

SOURCE: Ehrlich and Ehrlich (5), p. 6.

Urbanization*

At the present time about about 20 percent of the world's population lives in population centers greater than 100,000. By the year 2000, this figure is expected to increase to 40 percent. Urban regions presently cover about 10 percent of the U.S. land area, and this will increase to 16 percent by 2000 (17). Such a rush to urbanization is usually interpreted as a desire for a higher material standard of living, but the cost in the *quality* of living can be substantial. For example, the crime rate can jump from about 1000 crimes per year per 100,000 people in rural areas, to over 5000 crimes per year per 100,000 population for larger cities (17).

The close proximity of such masses of people also results in substantial pollution problems which may have been avoided for sparser settlements. In cities, heat from the sun is stored in the concrete and brick, creating an urban heat island which traps air pollutants and prevents

---

* Refer to Figures 2.1 and 2.2 for further information on this issue.

ventilation. Concentrating all of the human waste from a city and discharging this into a river usually overwhelms the ability of the river to assimilate the waste load without causing detrimental effects (assimilative capacity). Overtaxing the stream's assimilative capacity causes it to die an ugly and repulsive death. Noise, which would have been readily absorbed into the environment in rural areas, produces a cacophony that can result in irritation, mental duress, and physical harm.

Thus the pollutants produced by our society are amplified in their effect on the environment simply due to urbanization.

Consumption of Natural Resources

The third reason why we have pollution is that when we consume so many resources, we do it wastefully. We use non-replenishable materials (oil, metal ores, etc.) as if we had no need for them in but a few decades. Industries are not solely to blame for using new ("virgin") materials instead of recycled ("secondary") materials. The result of many existing government policies is to create economic incentives for using virgin materials over using secondary materials. Such policies as differential freight rates established by the Interstate Commerce Commission, and resource depletion allowances and other provisions of the federal tax code, result in benefits to the virgin-material-producing sectors of the economy as opposed to the secondary-material-production sectors (20). As an example, Table 7.2 shows the effects of tax benefits on recycling in the paper industry.

The United States, for example, uses almost one-half of all the world's natural resources, but represents only about six percent of the world's population. Other developed countries are only slightly less guilty.

A utopian dream, held by many, is that all of the poor countries in the world could eventually reach our level of wealth. Some day we may approach this dream, but certainly not under the present system of resource use, as is readily apparent. If we accept the goal of material wealth for all the world it must occur so that the materials and energy that are now consumed and wasted are either recycled or conserved— i.e., a balance is attained. The present system promotes pollution and encourages waste.

TABLE 7.2

COMPARISON OF VIRGIN MATERIAL TAX BENEFITS WITH VIRGIN AND SECONDARY MATERIAL PRODUCT COST DIFFERENTIAL

Paper product	Product cost*		Cost differential in favor of virgin material	Tax benefit as a percent of virgin and secondary material cost differential
	Using virgin material (dollars/ton)	Using secondary material (dollars/ton)		
Linerboard (100 percent virgin fiber compared with 25 percent secondary paper)	78.50	81.00	2.50	72
Corrugating medium (85 percent virgin [semi-chemical] compared with 35 percent secondary [semichemical])	79.50	82.00	2.50	72
Combination boxboard:				
100 percent virgin (kraft) compared with 100 percent secondary (newsback)	152.50	155.50	3.00	60
100 percent virgin (kraft) compared with 100 percent secondary (whiteback)	152.50	174.50	22.00	8
Printing and writing paper (100 percent virgin compared with 100 percent secondary)	92.00	99.00	7.00	26

*Cost at the point in processing where virgin and secondary materials are equivalent inputs.

SOURCES: Midwest Research Institute, "Economic Studies in Support of Policy Formation on Resource Recovery," unpublished report to the Council on Environmental Quality, 1972. Cost data modified from W. E. Franklin, *Paper Recycling; The Art of the Possible.* Washington, D.C.: American Paper Institute, 1973. Table taken from U.S. Environmental Protection Agency (20), p. 35.

Exploitation of Technological Skills

Finally, we as a species are just too clever for our own good. We can reason, and thus eradicate disease and eliminate misery. But this promotes pollution growth and consumption of resources—and the necessary production of all types of horrible wastes which must be absorbed into our ecosystem without creating undue stress. Chlorinated hydrocarbons, pesticides, polychlorinated biphenyls (PCB's), kepone, radiation, noise, and heat are but some of the many unwanted products of our skillful exploitation of resources. Without our technical and scientific skills, all devoted to the good of society, none of these pollutants would have existed. Without such a creature as the human, the world would not be subjected to such stresses upon its ecosystem.

7.3 MOVEMENTS FOR GLOBAL POLLUTION CONTROL

Lest the reader begin to view pollution as a local phenomenon, it might be wise to point out that many types of pollution are in fact global problems, which must be controlled by international treaties and agreements.

Some of these problems are bilateral, such as the controversy between the USA and Mexico over the salinity of the Colorado River. In this case, the river is used so extensively for irrigation purposes that it picks up a high saline level before crossing the border into Mexico. The high salinity makes its use for further irrigation impossible and the Mexican government is rightly concerned. The controversy has finally evolved to the agreement between the two countries that the river water would be treated in a huge desalinization plant before it flows into Mexico.

In addition to bilateral agreements, some treaties are among a small group of interested nations all of whom stand to gain from the cooperation. Examples of such agreements are the NATO environmental program in Europe and the European Economic Community research programs. The latter are organized by multinational committees, and research objectives and projects are assigned to various nations. This organization avoids duplication of effort and fosters cooperation and understanding. It should, however, be quickly pointed out that international cooperation in research can continue even in the midst of international pollution problems, as in the case of the acid-rain problem in Scandinavia discussed below.

The most difficult type of international attack on pollution problems is the global one, even though all nations (theoretically) have an interest and should therefore cooperate.

Unfortunately, there exists in the world great disparity in ethics, politics, and material wealth, and it is extremely difficult for all nations to agree on the seriousness of and approaches to the control of global environmental problems. Poorer nations, being less economically developed, quite logically interpret "environment" in terms of their own problems. They see poverty, bad housing, disease, and lack of adequate food as the principal threats to their "quality of life." And it is clear that as industrialization continues in these nations, leading to greater environmental problems, the people are quite willing to accept the degradation of their environment in order to enjoy the benefits gained by "development." As a result, these nations look with cynicism on the pious pronouncements of the rich nations toward curbing world pollution. They view the world environmental movement as just another attempt to perpetuate the colonial attitude and to "keep them in their place."

The breakdown of good will and agreements between nations can also occur among the more economically developed nations as well. A prime example of the absence of moral integrity occurs in the unwillingness of Japan and the USSR to adhere to the moratorium on the slaughter of whales. The other whaling nations, most notably Norway and the USA, agreed that a ten-year moratorium on whaling, drafted at the Stockholm Conference, should begin in 1970. This decision was especially difficult for Norway, which is a fishing nation and had one of the largest whaling fleets in the world. This small nation, however, was able to recognize the need to sacrifice economic gain for a global good. The decisions by Japan and the USSR, on the other hand, are telling indicators of their attitudes toward global cooperation and environmental protection.

The United Nations has been the main source of impetus toward attaining global agreements and cooperation in environmental problems. The Global Environmental Monitoring System has been initiated under the United Nations Environment Program (UNEP), and is a principal component of Earthwatch—a continuing data-gathering and analysis operation. UNEP's headquarters are in Nairobi, Kenya, where it has convened a number of conferences on various topics, all intended to foster international understanding of a common problem.

Unfortunately, narrow self-interest on the part of sovereign nations makes such cooperation a dream of the future.

7.4 TYPES OF POLLUTANTS

The unwanted residuals of our society can take many forms, and although they are not easily categorized, they can be described in one of several groupings. For example, one could speak of inert, tainting, and toxic substances, with the major concern obviously being health. We would like to describe pollutants more along engineering lines and speak of those that are normally discharged into water, into air, and onto land.

Water

From the earliest times, water has been used as a carrier of wastes. Interestingly enough, however, the first sewerage systems were for stormwater only, and disposal of wastes into the sewers was forbidden. Eventually, two drainage systems were constructed in most cities; a stormwater drainage system and a sanitary sewerage system.

Originally, the types of wastes introduced into the sanitary system were biodegradable—i.e., they could be readily decomposed by microorganisms. This decomposition can take place either in the presence of dissolved oxygen (aerobic) or in the absence of it (anaerobic). Since the end products of anaerobic decomposition, such as hydrogen sulfide, ammonia, etc., are considered foul, it is desirable to maintain aerobic conditions in watercourses. If the demand for oxygen is too great, the introduction of O_2 from the atmosphere and the production of it by algae in the water cannot keep up with the demand, and the water becomes anaerobic.

The maintenance of aerobic conditions therefore requires the control of the oxygen demand placed on a watercourse. The greater the organic load, the higher is the oxygen demand and the greater will be the probability of anaerobic conditions occurring. This demand is commonly referred to as BOD, Biochemical Oxygen Demand, and is an important measure of the effect of a waste on a watercourse.*

Typical BOD values of some common wastewaters are shown in Table 7.3. The relative ability of these waters to deplete oxygen levels is important, and such numbers serve to illustrate the impact of biodegradable *industrial* wastes on water quality.

Wastes that are biodegradable can have effects on watercourses other

---

*For a further discussion of BOD and water pollution, see references 9 and 16 in the Bibliography.

than oxygen depletion. Organic chemicals such as formaldehyde can be highly toxic at high concentrations, and literally pickle an entire stream, killing all life. At low concentrations, however, formaldehyde is readily decomposed by common microorganisms, and creates a substantial BOD.

The second broad class of water pollutants is the nonbiodegradable materials, generally introduced as industrial wastes. Mercury from chlor-alkali plants, for example, has received considerable attention as a pollutant, since it taints aquatic life and makes it unfit for consumption.

Heat is a pollutant that can cause distress in aquatic communities either directly (preventing fish spawning, for example) or indirectly, by encouraging diseases or reducing oxygen levels.

TABLE 7.3

TYPICAL BOD VALUES

	BOD_5 (mg/l*)
Natural stream or lake water	5
Raw domestic sewage	250
Milk processing waste	6000
Brewing and distilling waste	8000
Pulp and paper mill waste	10,000

*Milligrams of dissolved oxygen used per liter of water

Air

As with water, air has been used as a wastebasket for our wastes since caveman days. It wasn't until the Industrial Revolution, however, that organized communities produced enough airborne wastes in a sufficiently concentrated area to cause real concern.

This concern was not widespread, however, since the vast majority of people accepted dirty air as a necessary consequence of industrialization and employment. In the United States, true concern at the governmental level did not really begin until 1948, when an inversion condition covered the steel town of Donora, Pennsylvania, and the air became so foul as to kill twenty people and to cause nearly one-half of the fourteen thousand inhabitants to become ill.*

---

*For a detailed treatment of this subject, see reference 14.

Although smoke from coal combustion is historically the most important air pollutant, almost any material can be emitted from a chimney or tailpipe. Gases, solids, and liquids have all been shown to be detrimental. Mercury, for example, noted earlier as a major water pollutant, is also of concern when emitted from a number of different processes, including incineration.

On a national level, some materials have been recognized as especially dangerous, and the federal government has set ambient air quality standards for these pollutants.

Establishing ambient air quality standards is based upon determining what is called a threshold concentration. Exposure to pollutant concentrations above the threshold concentration will have a damaging effect. Exposure to concentrations below the threshold concentration will not have a damaging effect. Whether threshold concentrations really exist is a point of controversy. Several studies have indicated that they do not: if a compound is harmful at some concentration, then some harm will result from exposure to any concentration greater than zero. This is the assumption used by the Food and Drug Administration in banning from human consumption any food additive that has been shown to cause cancer in laboratory animals, even if the animals were treated with excessively high concentrations of the compound. Nonetheless, ambient air quality standards represent an approach to air pollution control currently in use in the United States.

Table 7.4 lists the six pollutants most commonly discussed and their sources and detrimental effects. In addition to these, several air pollutants have been classified as especially hazardous: beryllium, asbestos, vinyl chloride, and mercury. The approach taken to control these pollutants has been to develop specific control strategies applicable to the industrial processes that generate them. The strategies are continually upgraded as new information becomes available.

Table 7.5 lists the amounts of five pollutants emitted by several broad categories of sources. Note that ozone is not included in this table, but is included in Table 7.4. Very little ozone is emitted directly to the atmosphere. Ozone is formed in the atmosphere as a part of a complex series of reactions which generate photochemical smog.

Do not be misled by Table 7.5. The fact that transportation sources emit about one-half the total amount of air pollution does not mean they are the worst source of air pollution. One pound of carbon monoxide tends to be much less damaging than one pound of sulfur oxides. Several methods exist to weigh the emissions of the various pollutants and to combine them into an index that measures the damaging effects of the various sources. Treatment of these methods is beyond the scope of this chapter.

Sulfur oxides seem to be especially important as a health concern, and have been indicted as a major causative factor in all serious air pollutions incidents, such as that at Donora. The interaction of particulates (dis-

TABLE 7.4

MAJOR AIR POLLUTANTS

Pollutant	Source	Effect
Sulfur dioxide	Combustion of coal and oil	Damage to health, plants, and property
Carbon monoxide	Motor vehicles	Damage to health, plants
Nitrogen oxides	Combustion of gasoline, oil, and coal	Damage to health, plants
Hydrocarbons	Motor vehicles	Formation of photo-chemical smog
Ozone	Photochemical smog	Irritation; damage to health and plants
Particulates	Combustion; industrial processes	Damage to health and property

TABLE 7.5

NATIONWIDE AIR POLLUTION EMISSIONS,
BY POLLUTANT AND SOURCE (1974)
(in millions of tons per year)

Source	Particulates	Sulfur oxides.	Carbon monoxide	Hydrocarbons	Nitrogen oxides	Total
Transportation	1.3	0.8	73.5	12.8	10.7	99.1
Fuel combustion in stationary sources	5.9	24.3	0.9	1.7	11.0	43.8
Industrial processes	11.0	6.2	12.7	3.1	0.6	33.6
Solid waste disposal	0.5	0.0	2.4	0.6	0.1	3.6
Miscellaneous	0.8	0.1	5.1	12.2	0.1	18.3
Total	19.5	31.4	94.6	30.4	22.5	198.4

SOURCE: Council on Environmental Quality (7).

persed particles) with SO_2 illustrates one problem with air pollutants, that their interaction is sometimes unpredictable and usually results in additional damage. Sulfur oxides are highly soluble in water, and thus do not enter the deep reaches of the lungs simply because they are dissolved first in the mucus. Attached to a particle, however, sulfur oxides can get a free ride to the alveoli deep in the lungs, where gas transfer takes place, and once there, they can cause considerable damage.

Two types of waste energy are commonly associated with air pollution—radioactivity and noise. Radioactivity here refers to shortwave radiation resulting from the breaking apart of an atomic nucleus. Such radiation is harmful, and although not technically a constituent of air, we obtain our greatest fraction of extraneous radiation through air.

Radiation exposure is commonly measured in *r*oentgen *e*quivalents in *m*an, or *rem*, which measures not the amount of radioactivity, but its effect. Background radiation in the USA is between 100 and 250 millirems. It is estimated that with diagnostic x-rays, a person probably receives an annual dose of 200 rems on the average.

Since 100,000 rems is considered the minimum before damage to tissue occurs, and 500,000 is sufficient to cause death, the likelihood of damage due to background radiation is not great. Or so argues one school of thought.

A second opinion is that *any* amount of radiation above and beyond the natural background is potentially deleterious, i.e., there is no threshold. Historically, as we find out more and more about the effects of radiation, "safe thresholds" seem to get lower and lower. One wonders where the truth is.

A second type of waste energy usually transmitted through air is noise. Defined as "unwanted sound," this pollutant has unique attributes. It is simple to abate (stop the offending operation) and leaves no residual.

Noise is measured by a nonlinear (logarithmic) scale, decibels-A (dBA), which follows the response of the human ear to sound. Because it is not linear, care must be exercised in using the dBA scale. For example, a doubling of the strength of the sound is measured by an increase of 3 dBA. A tenfold increase in the strength of the sound is measured by an increase of 10 dBA (14). Silence, the perception of no sound at all, is arbitrarily defined as 0 dBA. Note also that the scale is not additive. For example, two sound sources each producing 85 dBA will together produce an overall sound of 88 dBA. The calculations required to manipulate measurements in dBA are not so straightforward as addition and multiplication.* Generally 85 dBA is considered loud enough to cause potential damage if the noise is prolonged, and 135 dBA is the threshold of pain. Table 7.6 lists some typical everyday sources of noise.

TABLE 7.6

TYPICAL NOISE LEVELS)
(in dBA)

Threshold of hearing	1
Normal breathing	10
Leaves rustling in breeze	20
Whispering	30
Quiet office	40
Homes	45
Quiet restaurant	50
Conversation	60
Automobile	70
Food blender	80
Niagara Falls at base	90
Heavy automobile traffic, or jet aircraft passing overhead	100
Jet aircraft taking off, or machine gun at close range	120

SOURCE: Ehrlich and Ehrlich (5), p. 177.

Land

Whatever waste isn't dumped into a watercourse or spewed up into the air must be deposited on the land. Included are outdated newspapers, glass bottles, metal cans, paper cups, plastic bottles, abandoned automobiles, demolition rubble, mine tailings, dead animals, fly ash, dewatered sewage sludge, and the garbage from our dining tables (21). The major sources of solid waste and the quantities generated are shown in Table 7.7. The most objectionable category of solid waste, and the one that is often the most difficult to manage, is that which originates in urban areas. The largest single class of urban solid waste, as shown in Table 7.8, is household and commercial refuse. (The typical composition of household refuse is given in Table 7.9.) Note, however, that of

---

*For a detailed treatment of this subject, see reference 14.

the nine categories listed in Table 7.8, the fourth largest amount generated is sludge from sewage treatment plants. The pollution that we remove to protect streams and rivers must be disposed of somewhere, and
that typically occurs onto the land.

Not only do water pollutants and air pollutants affect the land, but
quite often a land pollutant will eventually affect both air and water. A
municipal dump is a grand example of this interaction. As decomposition of the organics takes place, and rainwater and/or groundwater flows
through the refuse, various types of pollutants are leached out. The resulting wastewater, called leachate, has a BOD of about 10,000 mg/l,
and thus rates as a major water pollutant. Dumps also catch on fire,
either intentionally or otherwise, and the air pollution from such sources
is obvious. A burning dump in New Orleans, for example, used to be so
bad that wind shifts were used to predict the areas where hospitals would
be asked to assist an unusual number of asthmatics.

The problem of land pollution is intimately tied to the problem of
land use. The planning of our communities in such ways as to disallow
the use of land as a disposal site, or allow it only under strict conditions,
effectively reduces potential land pollution problems. There are many

TABLE 7.7

Major Sources of Solid Waste, 1967

| Source | Solid waste generated | | |
	Millions of tons per year		Percentage
Urban	256		7.0
Residential		128	3.5
Municipal		44	1.2
Commercial		84	2.3
Industrial	110		3.0
Agricultural	2,115		58.0
Plants		552	15.0
Animals		1,563	43.0
Mineral	1,126		30.8
Federal	43		1.2
TOTAL	3,650		100.0

SOURCE: Ad Hoc Group, Office of Science and Technology, *Solid Waste Management*. Washington, D.C.: U.S. Government Printing Office, 1969; p. 7. In Granville H. Sewell, *Environmental
Quality Management* (21), p. 227.

better alternatives to land disposal of our solid waste, including the recovery of materials and energy, or the reduction of this waste by changes in life style.

TABLE 7.8

AVERAGE SOLID WASTES COLLECTED IN 1958
FROM URBAN AREAS IN THE UNITED STATES

Category	Lbs./capita/day
Combined household and commercial refuse	4.29
Industrial refuse	1.90
Institutional refuse	0.16
Demolition and construction debris	0.72
Street and alley cleanings	0.25
Tree and landscaping refuse	0.18
Park and beach refuse	0.15
Catch basin refuse	0.04
Sewage treatment plant solids (sludge)	0.50
Total solid wastes collected	8.19

SOURCE: 1968 National Survey of Community Solid Waste Practices (HEW). In Granville H. Sewell, *Environmental Quality Management* (21), p. 228.

TABLE 7.9

TYPICAL COMPOSITION OF HOUSEHOLD REFUSE

Component	Percent of total	Pounds per person daily
Garbage	12	0.20
Paper	50	1.75
Wood	2	0.07
Cloth	2	0.07
Rubber and leather	2	0.08
Garden waste	9	0.31
Metals	8	0.28
Plastics	1	0.04
Ceramics and glass	7	0.24
Nonclassified	7	0.24
Totals	100%	3.50 lbs.

SOURCE: Based on 1967 analyses in Santa Clara County, California. In Granville H. Sewell, *Environmental Quality Management* (21), p. 229.

Interrelationships of Various Pollutants

It must be emphasized that pollution can not be easily categorized, and that some pollutants can cause simultaneous land, water, and air problems. One pollution problem is solved only at the expense of another. An excellent example is the removal of air pollutants (sulfur oxides and particulates) from the emissions of coal-fired electric power plants. Typically, a pollution-control device called a wet limestone scrubber would be installed to reduce the sulfur oxides and particulate emissions. However, the result is to transform an air pollution problem into a solid-waste management problem. A recent study (19) has shown that a 1,000 MW (megawatt: 1 MW = 1,000 kilowatts) coal-fired electric power plant would produce 50,000 pounds per hour of sulfur oxides and 100,000 pounds per hour of particulates if no pollution control devices of any kind were used. If a wet limestone scrubber were installed, the air pollution emission would be reduced to 5,000 pounds per hour of sulfur oxides (a 90 percent reduction) and 1,000 pounds per hour of particulates (a 99 percent reduction). Unfortunately, that is not the only result: the process produces about 300,000 pounds per hour of calcium sulfate sludge, a waste material that no one really wants, and which must be disposed of in large holes in the ground. In addition, operating the wet limestone scrubber uses about 40 MW (4 percent of the capacity of the plant) of power.

In domestic wastewater treatment, solids are removed from the water in order to make it clean, and these materials (sludge) become a solid-waste problem. Incineration to get rid of the solid waste, however, produces an air pollution problem, especially in the removal of heavy metals.

Another case, yet to be resolved, is the chlorination of wastewater treatment plant effluents. This is thought to be necessary so as to disinfect the water. Unfortunately, various compounds in the water will react with chlorine to form dangerous substances such as chloroform, a known carcinogen. Although the levels of such compounds have been found to be low, their presence casts doubt upon the wisdom of chlorinating effluents.

These examples emphasize the fact that decisions about pollution control will always involve environmental trade-offs. The best one can hope to achieve in pollution control is to transfer the pollutant from one environmental medium (air, water, or land) to another, and/or to transform the pollutant. The objective of these activities is to create a situation that is less environmentally damaging than the uncontrolled case.

If we agree that something must be done to solve our increasingly complex pollution problems, we must inquire as to what approaches are available for solving these problems. Basically, there are three—the technical response, the behavioral response, and the anarchist response. We will look at each of these in turn.

7.5 TECHNICAL RESPONSES TO POLLUTION

After recognizing what pollution is, and what types of pollutants are a problem, the solution-oriented individual is anxious to get on with the solutions. A technically oriented person will understandably seek technical solutions, and these solutions can vary as widely as the types of pollutants involved. Engineers' responsibilities include the selection of that series of solutions which, to them, best fits the problem. A "best solution" is a careful balance between economic constraints, environmental impact, physical requirements, and public perception. In this section only a limited number of solutions are introduced, and it must be remembered that these are used only as examples of the types of thinking that go into seeking technical solutions.

There are basically three strategies used to solve a pollution problem by technical means—(1) change the process that produces the waste, (2) dilute or disperse the waste sufficiently to avoid problems, or (3) treat the waste by removing constituents or changing it so as to make it more readily acceptable for disposal into the environment.

Change the Process

A rule of thumb in environmental engineering is that the least expensive way of solving a pollution problem is as far upstream in the process as possible. Many industrial waste problems have been solved at a fraction of treatment cost by simple and inexpensive changes in the manufacturing process, instead of treating the effluent from the existing process.

For example, an electroplating plant was having difficulties with treating a waste which had, among other things, heavy metals, the worst of which was hexavalent chromium. Treatment for the removal of this waste would require reduction to trivalent chromium and precipitation as a hydroxide at a high pH—an expensive proposition. Upon investigation, however, it was discovered that almost all of the chromium came from drippings on the floor, which were then flushed into the sewer. The solution to the problem involved the construction of a drip trough which directed the drippings back into the main tank and kept them off the floor. A million-dollar problem was solved for about twenty dollars. *

Another example of source correction is the change of fossil-fuel power stations from burning coal to burning oil or natural gas. Coal is a notoriously dirty fuel, with particulates (fly ash) and sulfur oxides being the major harmful air pollutants. Although newer furnace designs have greatly increased coal burning efficiency, the emissions are still sufficiently dirty to require correction. Since retrofit emission treatment devices are quite expensive, the power companies sought cheaper solutions. One such solution was to burn gas and oil instead of coal, and many stations switched over to those fuels when the emission control requirements were enforced. It is in fact quite difficult to make oil or gas boilers smoke, and sulfur levels are usually quite low, hence solving both problems.

Unfortunately, when the oil and natural-gas shortages hit, these same power companies began to blame the regulatory agencies for forcing them to switch away from coal. This was untrue, since the power companies *chose* the solution, recognizing the potential problems. What seemed to be a reasonable source-correction solution at one time turned out to be an unfortunate choice.

Dispersal and Dilution

The second possibility for achieving pollution control is to disperse the offending material. This option is often stated as, "the solution to pollution is dilution." This seems reasonable under our definition of pollution, since something is a pollutant only at a sufficiently high concentration to cause a problem. The danger with this approach is that although a *local* pollutant concentration problem may be averted, the *global* problem may persist.

Disposal of biodegradable wastes into a large river is a ready example of dispersal. If the organic load on any part of a river is too great, the water will not be able to maintain a sufficiently high concentration of dissolved oxygen, and will go anaerobic—an undesirable condition.

Natural waters are in effect complete purification systems. If an or-

---

*Personal communication from George E. Barnes, Emeritus Professor of Sanitary Engineering, University of North Carolina, Chapel Hill.

ganic material that can be used as a source of food by microorganisms in the river is deposited into the water, the microbes attack this material, assimilate it, lower its energy level, produce more microbes—and in the process use oxygen. This process is known as respiration.

If the load of decomposable organic matter is great (a high BOD) and the rate of microbial decomposition is rapid, it is possible to overwhelm the availability of oxygen, most of which must be dissolved into the water from the air. If this happens, all life forms that require dissolved oxygen (e.g., fish) die, and the stream becomes vile. Microorganisms that exist in the absence of oxygen (anaerobes) will inhabit such streams and continue to decompose the organic material, but in the process produce such odoriferous gases as hydrogen sulfide and ammonia. Only a few organisms, such as sludge worms, are able to survive. The streams are black in color, with floating sludge mats and bubbles rising to the top due to gas formation in the bottom sediments, and are generally unattractive.

If, however, the organic load is not so great as to produce oxygen depletion, the stream will soon be purified, even though the level of oxygen might fall below full saturation, as shown in Figure 7.1A. Such a dip in the oxygen level is not detrimental to aerobic aquatic life as long as the level does not go below about 5 mg/l of oxygen.

In large rivers, if an effluent is dumped at one shore, it is possible for a part of the river to go anaerobic, as shown in Figure 7.1B. This is an unacceptable situation, and can be avoided by proper dilution, by constructing a manifold into the river as shown in Figure 7.1C. If this is done, the river will achieve self-purification without the problem of anaerobiasis. Using the assimilative capacity of the river for treating organic wastes is an old technique, and it is difficult to argue that this dispersion is detrimental to our environment.

Dispersion as a method of pollution control is also used where air pollution is a problem. Given a situation like that shown in Figure 7.2, the variables that affect what quantity of the stack emissions actually reaches the ground can be grouped in the following manner:

$$C \text{ is a function of } \frac{Q, U}{H, V, T, X}$$

where C = ground-level concentration of the pollutant emitted; Q = total emission from the stack; U = wind velocity; H = stack height; V = exit velocity of the stack emission; T = temperature of emissions; and X =

FIGURE 7.1

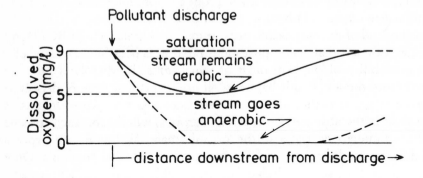

A. Dissolved oxygen level in a stream receiving biodegradable pollution.

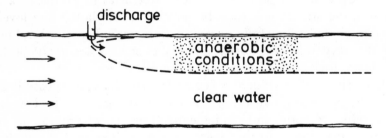

B. Without dispersion, the stream may be partially anaerobic.

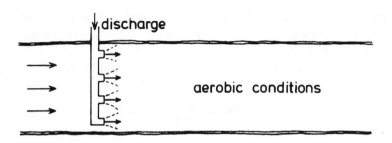

C. With dispersion, the demand for oxygen is spread out over the entire stream, thus avoiding anaerobic conditions.

downwind distance. There normally is one other variable in the equation—atmospheric stability. In our simplified case, we can say that the ground-level concentration C of a pollutant at any specified atmospheric condition is a function of the variables shown.

Obviously, there is little that can be done about wind velocity at any given location, although this variable can certainly be considered in the *siting* of a future source of pollution. The variables under control are the height of the stack, the velocity, and sometimes the temperature. Hence for a given Q, it is advantageous to have high values of H, T, and V in order to achieve low C values. And this is in fact done, with stack heights often exceeding 375 meters (1250 feet).

FIGURE 7.2

DISPERSION OF AIR POLLUTANTS

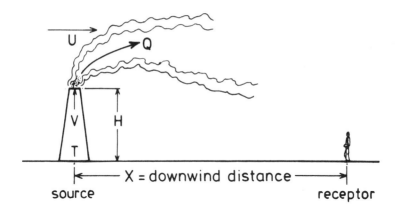

Unfortunately, this solution is acceptable only on a local scale, since the global effect can be highly detrimental. An example of this effect is the acid rain falling on Scandinavia. Sulfur oxides, mostly SO_2, are produced by the burning of fossil fuels as well as by some industrial processes. By building high stacks, the level of SO_2 in local areas, such as some of the industrialized regions in Europe, can be lowered to acceptable concentrations. When these sulfur compounds are injected into higher air currents, however, they are carried long distances. During this travel, especially over saltwater, sulfur dioxide is oxidized to sulfate aerosols such as sulfuric acid, which washes out with rain. The result can be devastating to lakes and forests in such fragile ecological systems as exist in Scandinavia. The mean pH of the rainfall in southern Norway, for example, is 4.3, whereas normal unpolluted rain would have a pH of about 5.0. This reduction in pH has changed the chemical composition of many Norwegian lakes and killed several species of fish. The salmon hatcheries have recently experienced great difficulty with mortality among newly hatched fry, due to the low pH of the water, and 111 lakes in southern Norway have lost all of their fish population due to the low pH (2).

This is not pollution control. Tall stacks, when they export a problem from one area of the world to another, are thus inappropriate methods of pollution control.

Treatment of Waste

The third method of pollution control is the treatment of the waste before discharge into the environment. Usually this is the most expensive alternative and should be considered only whenever all else fails.

For better or worse, most of our cities are constructed so that sanitary wastes (wastewater from toilets, kitchen, and washing) are collected in sewers and discharged at one or several main points. Years ago, when the quantity of these discharges was small compared to the assimilative capacity of the receiving waters, this system was thought to be acceptable. With increased populations and industrial loads on the watercourses, domestic wastewater soon became a serious problem. Perhaps, if we have been more forward-looking, other methods of domestic excreta disposal could have been developed, but now it is too late, and we are stuck with the problem of treating this wastewater.

The techniques of wastewater treatment are numerous and are the subject of many engineering textbooks. It is not possible, therefore, to cover the field here. We will attempt instead to discuss one aspect of municipal wastewater treatment—the removal and disposal of the *solids* in wastewater.

Although the per capita contribution of solids into domestic wastewater is substantial (150 grams per day), sewage is still quite dilute.* The concentration of solids is normally about 250 mg/l, meaning that the fraction of solid material that must be removed comprises only 0.025 percent of the total flow. When you consider that raw sewage is better

---

*We are limiting our discussion here to "suspended solids," or those solids that would filter out on a filter paper. Dissolved solids, although also a problem, are not considered in this discussion.

than 99.9 percent pure water, and that in terms of solids, the effluent from a wastewater treatment plant should be 99.998 percent pure, you might begin to appreciate that wastewater treatment is not a simple matter.'

It has been found, over many years of empirical experience, that these solids for the most part have densities different from the density of water and, given the opportunity, will sink or float.

Accordingly, "clarifiers" are employed to accomplish the removal of wastewater solids. Figure 7.3 is a simplified drawing of such a tank. The sewage enters the middle of the tank and is distributed radially outwards. The time that an average particle of sewage should spend in the tank (the theoretical detention time) determines how much of the solid material is removed before the sewage exits over a peripheral weir, a steel plate with V-notches in it to distribute the flow evenly over the entire tank. Practically, it is not possible to build large enough tanks even to approach 100 percent removal, and thus some compromise is reached. In most cases, the removal rate can be plotted against the detention time, as in Figure 7.4, with the normal detention time being two hours. Note that most clarifiers are only about 60 to 70 percent effective. Achieving greater removal efficiencies could only be provided at greater detention times. The increased settling-tank sizes that would be required would not be cost effective.

The solids that float to the top are skimmed off and buried. There

FIGURE 7.3

SIMPLE DIAGRAM OF A CLARIFIER USED IN WASTEWATER TREATMENT

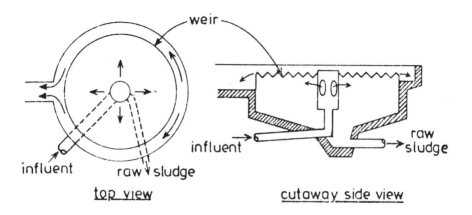

usually isn't very much of this grease, and thus it does not present further disposal problems.

The solid matter that settles to the bottom, on the other hand (now called "raw sludge"), has three important characteristics that make its further treatment and disposal difficult:

1. it is full of water—generally 95 percent water and only 5 percent solid;
2. it is rank, vile, odoriferous; and
3. the sludge contains pathogenic organisms.

Accordingly, further treatment is necessary before disposal into the environment is possible.

FIGURE 7.4

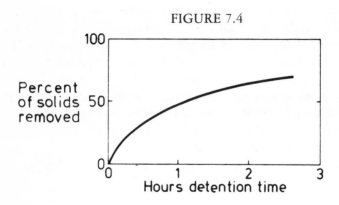

At this point it is interesting to note that there are two schools of thought on the disposal of this material. One school holds that this stuff is an embarrassing nuisance and that it should be "gotten rid of" in the quickest and cheapest way possible. The other opinion is that sludge is in fact a resource—it contains nutrients and organics which could be beneficial in some agricultural applications. The treatment alternatives are chosen according to the views held by the design engineers and their clients.

If sludge is considered a nuisance, dewatering (getting 80 percent of the water out by various mechanical means) and incineration is a logical sequence. This is a costly effort, almost always requiring supplemental fuel (oil or gas) and strict control of stack emissions. It is, however, a solution that reduces the problem to a pile of ash, and is therefore an appealing proposition for engineers and treatment-plant operators who seek neat, clean, and final solutions.

If the other strategy is accepted, of putting this material to some use, the three problems stated above must be addressed. Most of the water must be removed, simply because the sludge is otherwise too heavy for economical transport. If it is to be spread on land close to human habitation, the odor-causing organics must be stabilized to remove the unattractive quality of the material and, most importantly, the pathogenic organisms must be destroyed if the sludge is to come in contact with humans, either directly (e.g., on golf courses) or through agricultural products.

A standard procedure for attaining these ends is to digest and dewater the sludge. Digestion is accomplished in a large holding tank where microorganisms are allowed to decompose the organic matter. These organisms are anaerobic (they work in the absence of oxygen) and produce gases. In addition to ammonia, CO_2, and other gases, the microbes produce methane—CH_4, which is an excellent fuel and is used to heat the digester and increase the speed of the biological reaction. A good deal of work is currently under way to make the most of the methane generated by anaerobic digestion. The design for a sewage treatment plant to be constructed in Denver, Colorado, will use the methane anaerobic sludge digesters to run three 100 KW dual-fuel generators, to heat the digesters, and to incinerate screenings (8). The electricity produced should meet all plant needs.

Typical digester gas is about 60–70 percent (by volume) methane, and thus has a heat value of 600–650 Btu per cubic foot (British thermal unit: 1 kwh = 3,413 Btu), slightly more than one-half the heat value of natural gas, which is itself mostly methane. The normal rate of gas production in a sewage treatment plant is about one cubic foot per person served per day. A typical digester, shown in Figure 7.5, has a floating cover, which is necessary because of variable gas volumes in the tank.

FIGURE 7.5

DIGESTER USED
IN TREATING SOLIDS
PREVIOUSLY REMOVED
FROM WASTEWATER

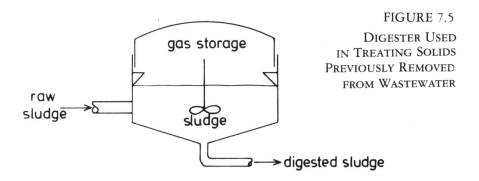

Although the nasty nature of the sludge is reduced considerably during its fifteen- to thirty-day detention in the digester, the pathogenic organisms have not all been destroyed, and the sludge is still potentially dangerous. Similarly, the digestion operation does little to increase the solids concentration (get rid of excess water) and dewatering is often necessary. In smaller plants, this dewatering is done by spreading the sludge over sand beds and allowing it to drain and dry. Since this process takes about sixty days at best, large land areas are required,.and large plants cannot afford such a luxury. Mechanical dewatering devices are therefore employed at large installations.

The dried, decomposed (but not sterile) sludge can then be spread on land, or further processed to make commercial soil conditioner. It is not a very high quality fertilizer, and it does have problems with unwanted heavy metals (e.g., zinc, lead, cadmium) which can be toxic to crops or to people who eat the crops. Sludge can also contain seeds that survive digestion. Tomato seeds seem to be especially hardy, and every pile of dried sludge will eventually sprout healthy tomato plants.

The sludge treatment process described above is not the only one available to the engineer. For example, pasteurizing of the sludge is practiced in Germany, and pyrolysis is used in the United States. In some areas the sludge is pumped out and sprayed onto fields and meadows in its raw state, with no water removed. Smaller towns have found that transport of digested sludge, without dewatering, is economical if the tank truck spreads it on fields. No health or other environmental problems have been encountered in these cases.

In short, the engineer has many options in designing the proper disposal scheme for wastewater solids. It is not a cookbook/handbook solution—it requires skills in human relations, natural sciences, and economics in addition to the standard engineering capabilities. In the environmental engineering area, there are no externalities.

7.6 BEHAVIORAL RESPONSES TO POLLUTION

Behavioral scientists (psychologists, sociologists, etc.) have for some years studied the problems of environmental degradation, and some have been bold enough to suggest that the solutions to our pollution problems are being approached in a backward manner. Since engineers have been in charge of pollution control, and engineers like nuts, bolts, and concrete, it is suggested that the solutions proposed thus far have been of a technical nature, and have not taken account of the behavior of people. They argue that engineers are like physicians in the nineteenth century, treating symptoms and not the disease, and that bloodletting and leeches are not the way of the future.

Instead of seeking technical solutions to pollution, the response of the behavioral scientists is to change the behavior of people so as to *eliminate* or reduce pollution at the source instead of simply treating it once it is created. An interesting argument, and difficult to refute on purely theoretical grounds. Practically, what are the types of solutions proposed?

One of the chief causes of pollution, as earlier noted, is overpopulation. If this arm of the problem is to be tackled, it seems reasonable to suggest that technical solutions (e.g., new birth-control methods) will not be sufficient, and a change in human values or ideas of morality must also occur (11). The solution to this problem involves informing humanity of the problem, and then asking each individual to contribute to its solution. A single individual, however, feels that he or she cannot at all matter—one newborn baby is after all only 1/4,500,000,000th of the problem—and as a result the total wrong is perpetuated. Indeed, all of human instincts point toward procreation, which at one time was simply a matter of survival. Ironically, with better medical care, increased food supplies, and advanced technology, our instincts may now well lead the human race toward destruction.

But how to change the drive to procreate? This has always been considered a sacred right of the family unit and nobody has ever had (yet) to seek permission to have a baby. Even the United Nations stated in 1968 that "any choice and decisions with regard to the size of the family must irrevocably rest with the family itself, and cannot be made by anyone else" (22). If we accept this argument, it follows that the only way to limit the growth of human population is to ask each family to limit the number of children.*

One means of doing this would be to appeal to conscience. Unfortunately, behavioral scientists have shown that such an appeal is psychologically pathogenic—one feels guilty if he accepts the pressures and succumbs to public opinion, since he has shown himself to be a weak simpleton, or he feels guilty because he has contravened such pressure. In short, we can show that "no good has ever come from feeling guilty, neither intelligence, policy or compassion" (10).

---

*For further discussion of this topic, refer to chapter 9.

If this then is not the answer, what is? This is the question that behavioral scientists are tackling by exploring ideas of personal and community space, overpopulation of a behavioral setting, and other concepts.

Research has thus far shown that some type of mutual coercion, according to Harden's well-read "Tragedy of the Commons" (11), is necessary. According to him, technical solutions are not possible, and "the only way we can preserve and nurture other and more precious freedoms is by relinquishing the freedom to breed."

This coercion is not an appeal to conscience, but a mutually agreed-on arrangement, much like our present traffic laws or societal conventions.

Of course, this entire question can be exploded by disagreeing with some of the original assumptions. How stable is the family unit? Is a family a practical institution in the twenty-first century? Would clones (test-tube people) be more practical, especially if their genetic characteristics can be carefully controlled? *

Such are the frightening questions we are asking our behavioral scientists to consider, and their answers will certainly influence our environmental problem of pollution.

Another, perhaps more acceptable behavioral response to pollution involves modifications to the standard of living currently enjoyed in the United States. That is, instead of limiting the number of people, one might approach the problem by limiting the amount of resources each one uses and the amount of pollution each generates. An excellent example of this is the familiar flush toilet, which uses about five gallons of water each time it is flushed. This quantity of water amounts to over forty percent of all in-house water use. Several alternatives to the flush toilet exist (13). Some are based on composting methods that use no water. Others are redesigns of the flush toilet which use substantially less water (two to three gallons per flush). Still others use waste water from washing clothes and dishes, instead of water of drinking quality. While the amount of human waste produced would remain constant, the net result of implementing these systems would be reductions in the volume of sewage to be treated and in the volume of water to be purified to drinking quality.

Energy consumption and pollution are closely tied. The less energy we consume, the less pollution we produce, under the current mechanisms for producing energy. Every mile not traveled by private auto-

---

*For further discussion on this issue, the reader is advised to refer to chapter 9

mobile, every degree lower the home-heating thermostat is set, every kilowatthour of electricity saved, reduces air pollution, water pollution, and land pollution.

7.7 ANTI-TECHNOLOGICAL RESPONSES TO POLLUTION

Years ago, in Boy Scout first-aid training classes, we were taught how to stop massive bleeding by the application of a tourniquet—a scarf tied tightly around an arm or leg; pressing on an artery near the surface. The worn-out joke was that the only place you would not put a tourniquet was around the neck.

Such "tourniquet around the neck" solutions to pollution control have, however, been suggested by proponents of a school of thought who might be labeled the "anti-technologists," but who seem to regard themselves as anarchists (1), in the sense of attaining the maximum freedom from the evils of modern technology. They look at the pollution problem as an outgrowth of our world order, and since our present world order is evil, one is morally obliged to do away with it, thus solving the pollution problem.

The anarchists save their sharpest barbs for capitalism. One reason for this is that the free capitalist system (democratic) is the only one that will tolerate such radical thought. The second reason is that capitalism is indeed a system where private ownership of resources leads to wide-scale destruction and despoilment. Further, capitalism promotes production and consumption, hence accelerating the rate of resource use. The anarchists believe that the "invisible hand" has not been able to temper such forces, and unchecked capitalism will eventually lead to economic disaster.

Because of their aversion to capitalism, these anarchists are often supported by ideologies that have as part of their dogma the missionary zeal of exporting their own economic systems, generally by force, to the remainder of the world.

Radical thinkers do however have a message that makes a great deal of sense to an increasing proportion of the world. Is it really necessary to create pollution by consuming energy and materials at such fantastic rates? Are three cars per family really necessary, where two bicycles might do? Are aerosol cans really required, when they are incompatible with our world ecology? Must hamburgers really be wrapped in a polystyrene box?

If all of the people began to ask some of these questions, which are not at all anti-technology but simply common sense, much of our pollution would never occur. The value of the anti-technology movement in pollution control is precisely that their overstatements of fact, when boiled down to the practical level, may eventually reduce the production of pollution.

7.8 CONCLUSIONS

It should be clear that pollution is not a problem without a solution, whether it be technical or behavioral, or a radical change of life style. In other words, it is possible to answer the "how" of pollution. But perhaps the "why" question is just as interesting.

If pollution control is a matter of survival, it should not be difficult to answer the "why." Unfortunately, even if we could prove that survival was at stake, the contribution of a single individual is minor and thus considered by him to be unimportant. In addition, the day of reckoning is too distant, and we are willing to enjoy short-term pleasures even if they lead to future disasters. After all, we'll all be retired or dead by that time, and why worry about posterity—what has it ever done for us?

If the "why" cannot be answered by self-interested fear, how can we then approach the problem?

One suggestion is to start at the beginning, with our religious inheritance, and work to the present, thus at least explaining our present predicament. White, among others, argues that the Judeo-Christian tradition is responsible for such western institutions as capitalism, urbanization, private ownership of resources, and hence pollution of the environment (25). We have suggested that perhaps White did not go back far enough, that it might have been a sense of values and morality that prompted the acceptance of the Judeo-Christian religions, and thus also the other components of western civilization (24). The famous passage in Genesis where man is given the right and duty to replenish the earth, subdue it, and have dominion over all living things, was after all man's interpretation of his role in the world. We cannot therefore blame our religions for the mess we are in. By the same token, we cannot ask them to lead us out of this predicament, especially since their dogma is unable to handle a crisis that was not even dreamed of when the traditions and liturgy were developed. A sad example of this is the Catholic church's continued resistance to birth control.

Obviously, our existing religions and moral strictures are useless, and a new theology is needed. René Dubos, among others, has proposed that we should develop a new theological approach toward the world, to view it not as nature to be conquered, but as a home in which to live (4). This new religion, with its own ethic, is still at its embryonic stage, but it is already clear that like all other religions, the theology of the earth requires a "feeling," or a faith, which cannot necessarily be explained by rational (scientific) means. Neither can this feeling be taught or casually distributed to the populace. It is a personal reawakening—occurring at unexpected moments—when the canoe glides through the quiet mist of a lake, or the view from the mountain top after a hard bike ride, or watching children swim in a previously polluted river.

7.9 EXERCISES

1. What does "pollution" mean to you? List some acceptable definitions and argue each in terms of strengths and weaknesses.

2. Do you agree with the claim that "the present system promotes pollution and encourages waste"? Why?

3. Write a short essay on water pollution and cite some examples, with references.

4. List ten of the most polluted (any form) regions of the world. Explain the reasons of pollution of at least five of them.

5. Do you find our present pollution control technologies sufficient? If not, what would you suggest? What is the weakest spot?

6. Study reference 4 in the Bibliography and write a short essay on it.

7. Study reference 11 and write a short essay on it.

8. Study reference 25 and write a short essay on it.

9. Do you believe that the United Nations can play an effective role in controlling pollution on a global level? Discuss.

10. Do you think man is capable of polluting outer space? Write a short essay on it.

7.10 BIBLIOGRAPHY

1. Bookchin, Murray. "Ecology and Revolutionary Thought," in *Post-Scarcity Anarchism.* New York: Times Change Press, 1970.
2. Braekke, F. H., ed. *Impact of Acid Precipitation on Forest and Freshwater Ecosystems in Norway.* Research Report FR6/76, Norwegian Institute for Water Research, Oslo.
3. Commoner, Barry. "The Ecological Facts of Life." Background paper prepared for the 13th National Conference of the U.S. National Commission for UNESCO, 1969.
4. Dubos, René. "A Theology of the Earth," in G. E. Frakes and C. Solberg, eds., *Pollution Papers.* Appleton-Century-Crofts, 1970.
5. Ehrlich, Paul R., and Anne H. Ehrlich. *Population. Resources and Environment.* Second edition. San Francisco: W. H. Freeman, 1970.
6. *Energy Facts.* Prepared for the Subcommittee on Energy of the Committee on Science and Astronautics, U.S. House of Representatives. Washington, D.C.: U.S. Government Printing Office, November 1973.
7. *Environmental Quality.* Sixth Annual Report of the Council on Environmental Quality, December 1975.
8. *Environmental Science and Technology.* vol. 11, January 1977, p. 15.
9. G. M. Fair, et al. *Water and Wastewater Engineering.* New York: Wiley, 1968.
10. Goodman, Paul. In *New York Review of Books.* May 23, 1968. Quoted in ref. 11.
11. Harden, G. "The Tragedy of the Commons." *Science.* vol. 162, December 13, 1968, pp. 1243–48.
12. Hardy, W., as quoted in J. T. Hardy, *Science. Technology and the Environment.* Philadelphia: W. B. Saunders, 1975.
13. Love, Sam. "An Idea in Need of Rethinking: The Flush Toilet." *Smithsonian.* vol. 6, May 1975, pp. 60–66.
14. Magrab, Edward B. *Environmental Noise Control.* New York: Wiley, 1975.

15. Maunder, W. J. "What Is Pollution?" in W. J. Maunder, ed., *Pollution.* Victoria, B.C.: University of Victoria, 1969.

16. Metcalf and Eddy. *Wastewater Engineering.* New York: McGraw-Hill, 1973.

17. Miller, G. T. *Living in the Environment.* Belmont, Calif.: Wadsworth, 1975.

18. Odum, E. P. *Fundamentals of Ecology.* Philadelphia: W. B. Saunders, 1971.

19. Reiquam, H. N. Dee, and P. Choi. "Assessing Cross-Media Impacts." *Environmental Science and Technology.* vol. 9, February 1975, pp. 118–20.

20. "Resource Recovery and Source Reduction," Second Report to Congress, U.S. Environmental Protection Agency. Washington, D.C.: U.S. Government Printing Office, 1974.

21. Sewell, Granville H. *Environmental Quality Management.* Englewood Cliffs, N.J.: Prentice-Hall, 1975.

22. Thant, U, in *International Planned Parenthood News.* February 1968. Quoted in ref. 11.

23. Turk, Amos, et al. *Environmental Science.* Philadelphia: W. B. Saunders, 1971.

24. Vesilind, P. A. *Environmental Pollution and Control.* Ann Arbor Science, 1975.

25. White, L. "The Historical Roots of Our Ecological Crisis." *Science.* vol. 155, March 10, 1967.

Chapter Eight

BIOMEDICAL ENGINEERING

8.1 WHAT IS BIOMEDICAL ENGINEERING?

Since the dawn of time medicine has been practiced in one form or another. Likewise, evidence of engineering can be found in the early pages of the Bible and recorded in early historical documents. However, until the twentieth century little had been done to effect a marriage of these two scientific disciplines. There are many examples of measurement made centuries ago that could be called applications of engineering to medicine. For example, in 1728 Hales first inserted a glass tube into the artery of a horse and crudely measured arterial pressure. Poiseuille substituted a mercury manometer for the piezometer tube of Hales, and Ludwig added a float and devised the *kymograph*, which allowed continuous permanent recording of the blood pressure. It is only recently that electronic systems using strain gages as transducers replaced the kymograph.

Many medical instruments were developed as early as the nineteenth century—for example, the electrocardiograph, first used by Einthoven at the end of that century. Progress was rather slow until after World War II, when a surplus of electronic equipment, such as amplifiers and recorders, became available. At that time many technicians and engineers, both within the industry and on their own, started to experiment with and modify existing equipment for medical use. This process occurred primarily during the 1950s and the results were disappointing, for the experimenters soon learned that physiological parameters are not measured in the same way as physical parameters. They also encountered a severe communication problem with the medical profession.

During the next decade many instrument manufacturers entered the field of medical instrumentation, but development costs were too high and the medical profession and hospital staffs were suspicious of the new equipment and often uncooperative. Many developments with excellent potential seemed to have become lost causes. It was during this period that some progressive companies decided that rather than modify existing hardware, they would design instrumentation specifically for medical use. Although it is true that many of the same components were used, the philosophy was changed; equipment analysis and design were applied directly to medical problems.

A large measure of help was provided by the U.S. government, in particular by NASA (National Aeronautics and Space Administration). The Mercury, Gemini, and Apollo programs needed accurate physiological monitoring for the astronauts; consequently, much research and development money went into this area. The aerospace medicine programs were expanded considerably, both within NASA facilities and through grants to universities and hospital research units. Some of the concepts and features of patient-monitoring systems presently used in hospitals throughout the world evolved from the base of astronaut monitoring. The use of adjunct fields, such as biotelemetry,* also finds some basis in the NASA programs.

In the 1960s, an awareness of the need for engineers and technicians to work with the medical profession developed. The major engineering technical societies recognized this need by forming "Engineering in Medicine and Biology" subgroups, and new societies were organized. Along with the medical research programs at the universities, a need developed for courses and curricula in biomedical engineering, and today almost every major university and college has some type of biomedical engineering program.

Other examples could be quoted from early history, but in the main, they would not contribute much to an understanding of the field. Stated simply, biomedical engineering is the application of engineering principles and practices to the living systems that encompass the science of biology and the field of medicine. It is sometimes called bioengineering,

---

*The measurement of biological variables remotely by the use of radio signals.

either as an all-encompassing term or as a special term.

The prefix *bio-* denotes something connected with life. Biophysics and biochemistry are relatively old "interdisciplines," in which basic sciences have been applied to living things. One school of thought subdivides bioengineering into different engineering areas—for example, biomechanics and bioelectronics. These categories usually indicate the use of that area of engineering applied to living rather than to physical components. *Bioinstrumentation* implies measurement of biological variables, and this field of measurement is often referred to as *biometrics*, although the latter term is also used for mathematical and statistical methods applied to biology.

Naturally committees have been formed to define these terms, and professional societies have become involved. The latter include the IEEE Engineering in Medicine and Biology Group, the ASME Biomechanical and Human Factors Division, the Instrument Society of America, and the American Institute of Aeronautics and Astronautics. Many new "cross-disciplinary" societies have also been formed.

A few years ago an engineering committee was formed to define bioengineering. This was Subcommittee B (Instrumentation) of the Engineers Joint Council Committee on Engineering Interactions with Biology and Medicine. Their recommendations are as follows:

Bioengineering is the application of the knowledge gained by a cross-fertilization of engineering and the biological sciences so that both will be more fully utilized for the benefit of man.

Bioengineering has at least six areas of application, which are defined below.

Medical Engineering is the application of engineering to medicine to provide replacement for damaged structures.

Environmental Health Engineering is the application of engineering principles to control the environment so that it will be healthful and safe.

Agricultural Engineering is the application of engineering principles to problems of biological production and to the external operations and environment that influence it.

Bionics is the study of the function and principles of operation of living systems with application of the knowledge gained to the design of physical systems.

Fermentation Engineering is engineering related to microscopic biological systems that are used to create new products by synthesis.

Human Factors Engineering is the application of engineering, physiology, and psychology to the optimization of the man-machine relationship.

More recently, as applications have emerged, the field has produced definitions describing the personnel who work in it. A tendency has been to define the biomedical engineer as a person working on research in the interface area of medicine and engineering, whereas the practitioner working with physicians and patients is called a *clinical engineer.*

One of the societies that has emerged in this interface area is the Association for the Advancement of Medical Instrumentation (AAMI). This association consists of both engineers and physicians. In late 1974, they came up with a definition which is a good description.

> A *clinical engineer* is a professional who brings to health care facilities a level of education, experience, and accomplishment which will enable him to responsibly, effectively, and safely manage and interface with medical devices, instruments, and systems and the use thereof during patient care, and who can, because of this level of competence, responsibly and directly serve the patient and physicians, nurses, and other health care professionals relative to their use of and other contact with medical instrumentation.

Most clinical engineers go into the profession through the engineering degree route, but many started out as physicists or physiologists. Most of them have at least a B.S. degree, and many of them have M.S. and Ph.D. degrees.

On the other hand, there is also a need for support personnel. A new definition was coined a couple of years ago: "the biomedical engineering technician" (BMET).

> A *biomedical enginering technician* (BMET) is an individual who is knowledgeable about the theory of operation, the underlying physiological principles, and the practical, safe clinical application of biomedical equipment. His capabilities may include installation, calibration, inspection, preventive maintenance, and repair of general biomedical and related technical equipment as well as operation or supervision of equipment control, safety and maintenance programs and systems.

This was also an AAMI definition. Typically, the BMET has two years of pre-professional education at a community college. This person is not to be confused with the *medical technologist.* The latter is usually used in an operative sense, for example in blood chemistry and in the taking of electrocardiograms. The level of sophistication of a BMET is usually higher than the technologist's in terms of equipment, but lower in terms of the life sciences.

These definitions are all noteworthy, but whatever the name, this age of the marriage of engineering to medicine and biology is destined to benefit all concerned. Improved communication among engineers, technicians, and doctors; better and more accurate instrumentation to measure vital physiological parameters; and the development of interdisciplinary tools to help fight the effects of body malfunctions and diseases, are all a part of this new field.

Another major problem of biomedical engineering involves communication between the engineer and the medical profession. The language and "jargon" of the physician are quite different from those of the engineer. In some cases, the same word is used by both disciplines, but with entirely different meanings. Although it is important for the physician to understand enough engineering terminology to allow him to discuss problems with the engineer, the burden of bridging the communication gap usually falls on the latter. The result is that the engineer, or technician, must learn the doctor's language, as well as some anatomy and physiology, in order that the two disciplines can work effectively together.

The life scientists and the physician have been weak in quantitative approaches, whereas the engineer has quantitative methods as a real skill. Therefore, engineers can build up on the qualitative, empirical approach of the biologists, etc., while capitalizing on their descriptive skills in the field that the engineer may not have.

In addition to the language problem, other differences may affect communication between the engineer or technician and the doctor. Since the physician is often self-employed, whereas the engineer is usually salaried, a different concept of the fiscal approach exists. Thus, some physicians are reluctant to consider engineers as professionals and would tend to place them in a subservient position rather than as equals.

The probability is great that the 1970s will be known as the decade in which most rapid progress was made in this highly important field. There is one vital advantage that biomedical engineering has over many of the other fields that preceded it: the fact that it is aimed at keeping people healthy and helping to cure them when they are ill. Thus it may escape many of the criticisms aimed at progress and technology. Many purists have stated that technology is an evil. Admittedly, although the industrial age introduced many new comforts, conveniences, and methods of transportation, it also generated many problems. These include air and water pollution, death by transportation accidents, and the production of such weapons of destruction as guided missiles and nuclear bombs. However, even though biomedical engineering is not apt to be criticized as much for producing evils, some new problems have been created, such as shock hazards in the use of electrical instruments in the hospital. Yet these side effects are minor compared to the benefits that mankind can derive from it.

Within the practice of biomedical engineering, there are many sub-definitions. Some practitioners put biomedical or clinical engineers into specialists' categories similar to those in medicine. That is, they specialize in cardiology, neurology, etc., or they specialize by physiological systems such as the cardiovascular, nervous, or respiratory systems, etc.

The body can be considered as a number of engineering systems and the arts and sciences of engineering can be shown to have analogs in the body. For example, fluid mechanics normally associated with air or water flow becomes hemodynamics when applied to the blood. Kinematics, the mechanics of movement, applies equally to the limbs and to machine linkages. Human heat transfer and body heat regulation have their counterparts in the thermal behavior of physical systems. Strength of materials applies to bones equally with timber and steel, and the body is a fantastic automatic control system, with breathing, pumping of blood, and temperature being good examples.

8.2 CURRENT DIRECTIONS

This marriage of engineering and medicine is manifested in many ways. The major categories are:
1. Diagnostics—use of instrumentation to measure body signals.
2. Therapeutics—use of engineering devices to help cure or correct body ailments.
3. Prosthetics and orthotics—use of devices to repair, replace, or assist limbs, organs, and body parts.

These will be considered in turn with the use of some typical examples.

Diagnostics

The body can be diagnosed electrically, mechanically, chemically, thermally, and by radiation, but to discuss all of these would take more than one chapter in a book. To illustrate, let us take some very common examples, and look at the so-called *"man-instrument system."*

Figure 8.1 illustrates a man-instrument system. Not all the elements shown are always present, but they may be. First, the patient may have to be subjected to a stimulus to get a measurement. Perhaps this would be a light, a sound, or an electric shock. Most measurements, however, do not need a stimulus.

FIGURE 8.1

BLOCK DIAGRAM OF A TYPICAL MEDICAL INSTRUMENT

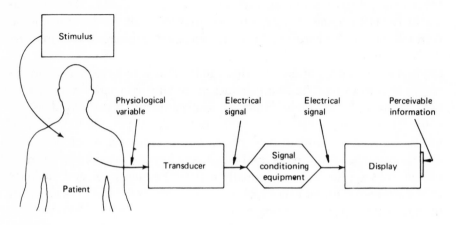

The *transducer* can be defined as a device capable of converting one form of energy or signal to another. In the man-instrument system, each transducer is used to produce an electric signal that is an analog of the phenomenon being measured. The transducer may measure temperature, pressure, flow, or any of the other variables that can be found in the body, but its output is always an eletric signal. Two or more transducers may be used simultaneously to obtain relative variations between phenomena. The transducer may be just a simple electrode.

The part of the instrumentation system that amplifies, modifies, or in any other way changes the electric output of the transducer is called *signal-conditioning* (or sometimes signal-processing) equipment. Signal-conditioning equipment is also used to combine or relate the outputs of two or more transducers.

In order to be meaningful, the electrical output of the signal-conditioning equipment must be converted into a form that can be perceived by one of man's senses and that can convey the information obtained by the measurement in a meaningful way. The input to the display device is the modified electric signal from the signal-conditioning equipment. Its output is one form of visual, audible, or possibly tactile information. In the man-instrumentation system, the display equipment may include a graphic pen recorder that produces a permanent record of the data.

Sometimes a control loop is needed, such as in biofeedback, described later in the chapter.

There are three common methods of diagnosis. The first makes use of direct body signals, the second uses conversion transducers, and the third employs radiation.

Direct Body Potentials. A direct biopotential is a signal that is generated by the body itself. The electrical potentials developed by the heart are known as electrocardiographs (ECG or EKG; the *K* is because the original definition was in German and was spelled with a *K*). The brain waves are known as electroencephalograms (EEG), and potentials developed by the muscles are known as electromyograms (EMG).

Figure 8.2 shows an electrocardiogram. The physician is familiar with what a normal wave looks like, and if he sees abnormalities, he can diagnose the medical problem. Each part of the wave has a letter designation (P, Q, R, S, T) and the time intervals between them or the time that part of the wave lasts is important. For example, a normal P-R interval is 0.12 to 0.20 seconds. The P wave takes about 0.11 second, and the S-T segment is 0.05 to 0.15 seconds, all for a normal heart beat. The length of the whole wave is one heart beat, and the normal ranges lie from 60 to 100 beats per minute. Also, the height (amplitude) and shape of the wave may change and in itself be significant.

FIGURE 8.2

ELECTROCARDIOGRAM SHOWING STANDARDIZATION PULSE

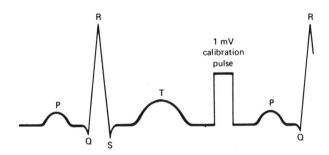

The measurements are made by means of an electrocardiograph. Electrodes are pasted onto the body in various known positions (the arms, legs, and chest). A wire from each electrode is connected to the machine. The signal obtained is amplified and displayed on a recorder.

Use of Transducers. If direct signals are not available in the body, a method must be devised to apply a signal that will change as body parameters change. These devices are commonly called transducers. For example, in critical-care units, it is necessary to obtain blood-pressure readings from inside an artery or a vein. A tube, known as a catheter, can be inserted into the artery or vein and it will "sense" the pressure. If the tube contains a sterile saline solution, the blood will press on the fluid and the fluid will relay the pressure to a point outside the body to the transducer. The transducer is actually an electrical resistance bridge, and as the pressure is applied, the resistance will change. If a voltage is applied to the resistance bridge, any changes in pressure result in resistance changes. This in turn will change the output signal, which when suitably amplified can be displayed on a cathode-ray tube or recorded. Thus, changes in blood pressure can be obtained as electrical measurements. This can be done in similar fashion with blood flow, temperature, and a variety of other physiological parameters. Figure 8.3 shows a catheter placed in a body.

FIGURE 8.3

CATHETERIZATION FOR CENTRAL VENOUS PRESSURE

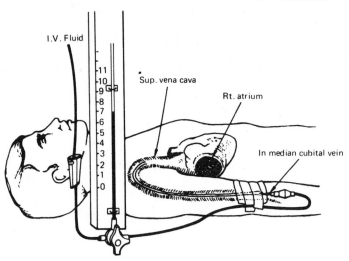

Use of Radiation Techniques. Another familiar diagnostic tool is the x ray. In 1895 Wilhelm Konrad Röntgen, a German physicist, discovered a previously unknown type of radiation while experimenting with gas-discharge tubes. He found that this type of radiation could actually penetrate opaque objects and provide an image of their inner structure. Because of this mysterious property, he called his discovery x rays. In many countries x rays are referred to as *Röntgen rays.* One year after Röntgen's discovery, Henri Becquerel, the French physicist, found a similar type of radiation emanating from samples of uranium ore. Two of his students, Pierre and Marie Curie, traced this radiation to a previously unknown element in the ore, to which they gave the name *radium,* from the Latin word *radius,* "ray." The process by which radium and certain other elements emit radiation is called *radioactive decay,* whereas the property of emitting radiation is called *radioactivity.*

The use of x rays as a diagnostic tool is based on the fact that various components of the body have different densities for the rays. When x rays from a point source penetrate a body section, the internal structure of the body absorbs varying amounts of the radiation. The radiation that leaves the body, therefore, has a spatial intensity variation that is an image of the internal structure of the body. When, as shown in Figure 8.4, this intensity distribution is visualized by a suitable device, a shadow image is generated that corresponds to the x-ray density of the organs in

FIGURE 8.4

USE OF X RAYS TO VISUALIZE THE INNER STRUCTURE OF THE BODY

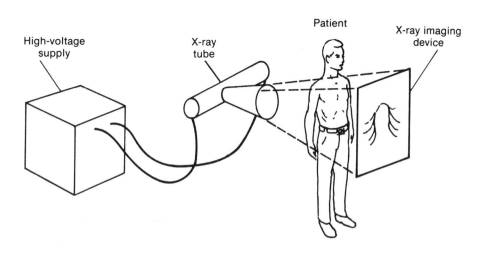

the body section. Bones and foreign bodies, especially metallic ones, and air-filled cavities show up well on these images because they have a much higher or a much lower density than the surrounding tissue. Most body organs, however, differ very little in density and do not show up well on the x-ray image. For some organs (stomach, colon, etc.), the contrast can be improved by filling the organ with a contrast medium (barium sulfate). The ventricles of the brain can be made visible by filling them with air (*pneumoencephalography*). Blood vessels can also be made visible by injecting a mass of a special contrast fluid (an iodized organic compound) before the x-ray exposure (*angiography*).

When an image does not change rapidly, photographic film is used extensively to visualize the x-ray distribution.

Since radiation can be dangerous if used frequently, another technique using ultrasound has been developed where bones and organs can be measured by being bombarded with high-frequency sound waves, which are relatively harmless at low energy levels.

Therapeutics

Another application of engineering to medicine is for therapeutic and curative purposes. Examples such as heat, infrared, and electric shock are well known. Ultrasound has also become very popular. The sound waves literally massage tissues deep down, in muscles for example. However, a very dramatic method is comparatively new and is known as *defibrillation*.

The heart can lose its synchronism and go into a rapid irregular twitching of the muscular wall known as *fibrillation*. Until a few years ago, patients frequently died from this condition. Now they can be saved.

Essentially, the defibrillator consists of two large paddles that are placed across the patient's chest, through which a high-voltage, short-duration shock is given. If done correctly, the heart is defibrillated and the patient is saved.

Prosthetics and Orthotics

A further application is in the area of prosthetics and orthotics. Artificial arms and legs have been used for many years, but recent advances have made them relatively efficient. In addition to limbs, there have been many types of artificial organs developed in the last few years. The artificial kidney or dialysis machine is a good example. The kidney filters and purifies the blood. By drawing the blood out of the body using plastic tubes, it can be passed through a filtering system that emulates the real kidney and cleanses the blood for a few days. However, the patient has to go back to the machine continually. Many advances have been made. The first machines filled a whole room, while today they are relatively portable.

Experimentally, there have been artificial hearts and artificial livers, and it is conceivable that engineering design can produce almost anything. All it takes is time, money, and patience. However, the body is a wonderful device, and the engineer needs a whole room full of equipment to replace an organ a few inches wide.

Other advances have been with parts of the body—many advances in dentistry, artificial hips, and assist devices for damaged muscles, to name a few. The heart valve is another good example of engineering. Many patients have had many years added to their lives by having an artificial heart valve surgically implanted when their own valve collapsed or decayed.

However, in order to implant an artificial valve, it is necessary to perform so-called "open-heart" surgery. Open-heart surgery was only made possible by the invention of the heart-lung machine. The operating principle of this machine duplicates the real body system. Essentially, deoxygenated venous blood is taken out of a vein and pumped through the machine, which gives the blood its oxygen while maintaining blood temperature correctly. The oxygenated blood is then fed back into the circulatory system. In this way, the heart is not needed, so the surgeon stops the heart by an injection, performs the surgery, and then starts it again with an electrical shock afterwards. In the meantime, all the body organs receive their oxygenated blood as if the heart were operating. This is frequently called a heart-lung bypass.

Another assist device worthy of mention is the pacemaker. The natural heart beats because of a node inside which causes an electrical signal to contract the muscle of the heart and so causes the beat and pumping action.

Like the natural pacemaker, a device capable of generating artificial pacing signals and delivering them to the heart is also called a *pacemaker*. Artificial pacemakers come in a variety of forms. They can be internally implanted for use by patients with permanent heart blocks who may require assisted pacing for the rest of their lives, or they may be worn externally for temporary requirements.

Internal pacemakers are surgically implanted beneath the skin, usually in the region of the abdomen. Internal leads connect to electrodes that are inserted directly into the myocardium. Since there are no external connections for applying power, the unit must be completely self-contained with a power source capable of continuously operating the pacemaker for a period of years. Figure 8.5 shows the pacemaker electrodes surgically connected to the heart, and the placement of the pacemaker.

External pacemakers, which include all types of pacing units located outside the body, may have electrodes attached directly to the myocardium with the connecting wires penetrating the skin, or the electrodes may be introduced into one of the chambers of the heart through

FIGURE 8.5

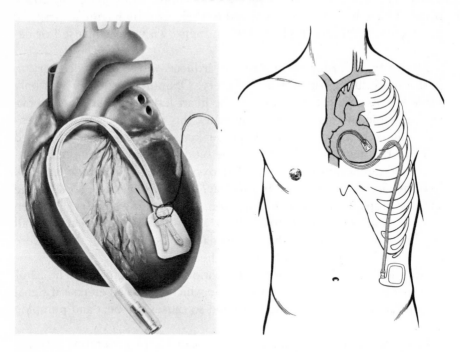

Left. pacemaker electrodes attached to the heart. *Right.* electrodes with pulse generator implanted in the abdomen. Courtesy of General Electric Company, Medical Systems Department, Milwaukee, Wisconsin.

Courtesy of General Electric Company, Medical Systems Department, Milwaukee, Wisconsin.

a cardiac catheter, called the *pacing catheter*. When the pacing catheter is used, the leads are brought out through the catheter. Most external pacemakers in use today are small portable battery-operated units. The pacemaker can be worn by an ambulatory patient, or it can be attached to the bed or arm of a patient in bed.

Again, many patients are leading long and useful lives because of a pacemaker. One present problem is that the batteries have to be replaced after two or three years of use, but as battery technology gets better, this will change. Nuclear-powered pacemakers have also been used.

Another application, although not quite as obvious, is the interface between engineering and psychology. The science of psychology is based on the nervous system and is interrelated with study of the senses. These two fields are often referred to as *psychophysics* and *psychophysiology*. There is no field of psychoengineering that has been defined as yet, but who knows for the future?

Electroencephalographs (EEG) and electromyographs (EMG) have been mentioned. These instruments measure the brain waves and muscle potentials respectively. Their use in psychological measurements is useful, since psychological changes affect the waves in various ways. One application of this is called *biofeedback*. The patient learns to recognize what a normal wave looks like, and when he experiences an abnormality, he is able to recognize it. If he is taught by an expert what to do, he can compensate and perhaps care for himself. The term biofeedback simply means that the information is fed back to the patient in order for him to try to correct it. An analogy is the familiar wall thermostat. If you want a room to remain at, say, 70° Fahrenheit, you set the thermostat at this level. If the temperature drops, a sensing device in the thermostat causes a switch to activate and turn on a heater or furnace. When the set temperature is reached, this is sensed and the furnace is turned off.

Patients have learned to recognize signals and have been able to cure headaches, bad eating habits, and even diseases like diabetes. In the latter case, the patient learns to control the secretion of his own insulin.

Psychological analysis has also been done by using *biotelemetry*. The word means making biological measurements over a distance. That is, the body signals are collected and are transmitted by a radio transmitter just as speech is transmitted by radio stations and by persons with "walkie-talkies." Behavior patterns can be studied both for research and clinically.

Telemetry is also being used for transmission of the electroencephalogram. Most applications have been involved with experimental animals for research purposes. One example is in the space biology program in the Brain Research Institute at the University of California, Los Angeles, where chimpanzees have had the necessary EEG electrodes implanted in the brain. The leads from these electrodes are brought to a small transmitter. Other groups have developed special helmets with surface electrodes for this application. Similar helmets have been used for the collection of EEGs of football players during a game.

Telemetry of EEG signals has also been used in studies of mentally disturbed children. The child wears a specially designed "football helmet" or "spaceman's helmet" with built-in electrodes so that the EEG can be monitored without traumatic difficulties during play. In one clinic the children are left to play with other children in a normal nursery school environment. They are monitored continuously while data are recorded.

One advantage of monitoring by telemetry is to circumvent a problem that often hampers medical diagnosis. Patients frequently experience pains, aches, or other symptoms that give trouble for days, only to have them disappear just before and during a medical examination. Many insidious symptoms behave in this way. With telemetry and long-term monitoring, the cause of these symptoms may be detected when they occur or, if recorded on magnetic tape, can be analyzed later.

Lie detectors may not seem like an example of biomedical engineering, but they are. The so-called polygraph* depends partially on the *galvanic skin response*. It turns out that the electrical resistance of the skin varies considerably with the amount of perspiration generated. The sweat glands become overactive when a person is nervous such as when telling a deliberate lie. The palms of the hand are particularly susceptible to this, and so if an electrode is placed on the palm and another on the thumb, the electrical resistance changes with the perspiration, which is a measure of nervousness. This measurement can be observed, recorded, and analyzed.

8.3 ENGINEERING IN THE HOSPITAL

A complete analysis of a hospital would be too complicated for a work of this nature, but one area of biomedical instrumentation that is becoming

---

*Example: "Manual for the Safe Use of Electricity in Hospitals," 76-B-T Boston, National Fire Protection Association (NFPA), 470 Atlantic Avenue; 1973.

increasingly familiar to the general public is that of patient monitoring. Here electronic equipment provides a continuous watch over the vital characteristics and parameters of the critically ill. In the coronary-care and other intensive-care units in hospitals, thousands of lives have been saved in recent years because of the careful and accurate monitoring afforded by this equipment. Public awareness of this type of instrumentation has also been greatly increased by its frequent portrayal in television programs, both factual and fictional.

In hospitals that have engineering or electronics departments, patient-monitoring units, both fixed and portable, form a substantial part of the workload of the engineers and technicians, who are usually involved in the design of facilities for coronary and other intensive-care units and work closely with the medical staff to ensure that the equipment to be installed meets the needs of that particular hospital. Ensuring the safety of patients who may have conductive catheters or other direct electrical connections to their hearts is another function of these biomedical engineers and technicians. They also work with the contractors in the installation of the monitoring equipment and, when this job is completed, supervise equipment operation and maintenance. In addition, the engineering staff participates in the planning of improvements and additions, for there are many cases in which an intensive-care unit, even after careful design and installation, fails in some respect to meet the special needs of the hospital, and an in-house solution is required.

The need for intensive-care and patient monitoring has been recognized for centuries. The twenty-four-hour nurse for the critically ill patient has, over the years, become a familiar part of the hospital scene. But only in the last few years has equipment been designed and manufactured that is reliable enough and sufficiently accurate to be used extensively for patient monitoring. The nurse is still there, but her role has changed somewhat, for she now has powerful tools at her disposal for acquiring and assimilating information about the patients under her care. Thus she is able to render better service to a larger number of patients and is better able to react promptly and properly to an emergency situation. With the capability of providing an immediate alarm in the event of certain abnormalities in the behavior of a patient's heart, monitoring equipment makes it possible to summon a physician or nurse in time to administer emergency aid, often before permanent damage can occur. With prompt warning and by providing such information as the electrocardiogram record just prior to, during, and after the onset of cardiac difficulty, the monitoring systems enable the physician to give the patient the correct drug rapidly. In some cases, even this process can be automated.

Physicians do not always agree among themselves as to which physiological parameters should be monitored. The number of parameters monitored must be carefully weighed against the cost, complexity, and reliability of the equipment. There are, however, certain parameters that provide vital information and that can be reliably measured at relatively low cost. For example, nearly all cardiac-monitoring units continuously measure the electrocardiogram, from which the heart rate is easily derived. The electrocardiogram waveform is usually displayed and often recorded. Temperature is also frequently monitored. On the other hand, there are some variables, such as blood pressure, in which the benefit of continuous monitoring is debatable in light of the problems associated with obtaining the measurement. Since continuous, direct, blood-pressure monitoring requires catheterization of the patient, the traumatic experience of being catheterized may be more harmful to the patient than the lack of continuous pressure information. In fact, intermittent blood-pressure measurements by means of the sphygmomanometer, either manual or automatic, might well provide adequate blood-pressure information for most purposes.

Since patient-monitoring equipment is usually specified as a system, each manufacturer and each hospital staff has its own ideas as to what should be included in the unit. Thus a wide variety of configurations can be found in hospitals and in the manufacturers' literature. Since cardiac monitoring is the most extensively used type of patient monitoring today, it provides an appropriate example to illustrate the more general topic of patient monitoring.

The concept of intensive coronary care had little practicality until the development of electronic equipment that was capable of reliably measuring and displaying the electrical activity of the heart on a continuous basis. With such cardiac monitors, instant detection of potentially fatal arrhythmias (losses of heart rhythm) finally became feasible. Combined with stimulatory equipment to reactivate the heart in the event of such an arrhythmia, a full system of equipment to prevent sudden death in such cases is now available.

In the intensive coronary care area, monitoring equipment is installed beside the bed of each patient to measure and display the electrocardiogram, heart rate, and other parameters being monitored from that patient. In addition, information from several bedside stations is usually displayed on a central console at the nurses' station.

As might be expected, many different room and facility layouts for intensive coronary-care units are in use. One type that is quite popular is a U-shaped design in which six or eight cubicles or rooms with glass windows surround the nurses' central monitoring station. Although the optimum number of stations per central console has not been established, a group of six or eight seems most efficient. For larger hospitals monitoring of sixteen to twenty-four beds can be accomplished by two or three central stations. The exact number depends on the individual hospital, its procedures, and the physical layout of the patient-care area. In certain areas in which recruitment of trained nurses is difficult, this factor could also be considered in selecting the best design.

There are clinical laboratories where blood chemistry is recorded automatically and analyzed on a computer. There are complicated therapeutic devices. It is estimated that maybe ten thousand different biomedical devices, instruments, and systems are available to hospitals today for use in patient care. Complex instrumentation is found in nearly every hospital diagnostic and treatment area. Mechanical, electrical, electronic, pneumatic, and hybrid devices are available to record body functions, the state of the patient, and the extent of his illness; to assume some of his human functions should they fail; and to assist in his care, from raising the foot of his bed to controlling the risk of infection.

Today there is a proliferation of electrical and electronic equipment in the hospital. If such equipment is in good working order and used properly, it is relatively safe. Occasionally, however, situations develop in hospitals where equipment is faulty or is used in a way that could precipitate an accident. Sometimes the personnel using the equipment are not aware of the potential hazard, which in most cases can be avoided with little effort. Thus it is important that all hospital staff be aware of safety regulations and precautions.*

In recent years, surgical and intensive-care procedures have necessitated the invasion of the body by wires, fluid-filled tubes, and the like. These are conductors of electricity. If any part of the patient's body touches a ground point, he will form a complete circuit and can cause electric current to flow. On the body surface, these currents would be harmless since the skin can handle milliamperes of current, but inside the body this is not so. The heart can be put into fibrillation by a few microamperes. This is known as *microshock* as distinct from *macroshock*

---

*This is not considered to be a very accurate device since some people secrete little perspiration.

where there is no direct connection to the heart. Such a patient is called *electrically susceptible.* One remedy has been the very careful grounding of hospital electrical systems, and this has been one of the solutions by which the application of biomedical engineering has saved many lives.

Before leaving the subject of hospitals, one recent innovation should be mentioned that is usually considered a part of the biomedical engineering function. That is the role of the computer.

In Chapter 6, the reader was exposed to the computer. The computer in medicine is a special application. In addition to bookkeeping, billing, and housekeeping, the computer is now found used in many ways. These would include:

1. *Data acquisition*—The reading of the instruments and transcribing of data can be done automatically under the control of the computer.

2. *Storage and retrieval*—In the modern hospital, admission and discharge information, physicians' reports, laboratory test results, and pharmacy records among other things are stored and retrieved on command.

3. *Assimilation and Organization of Information*—To generate a patient data file, the computer must assimilate all information from each patient as it arrives and file it with data already in the computer for that patient. The computer can then organize the data and prepare a report of all tests performed.

4. *Data Reduction and Transformation*—Sometimes the data resulting from digitizing an analog physiological signal such as an ECG or EEG would be quite useless if retrieved from a computer in raw form. To obtain information from such data, some form of reduction or transformation is necessary. Programs are now available to do this and make statistical and other analyses.

5. *Mathematical Operations*—If the physiological parameter cannot be measured directly, it may be calculated from other variables that are readily accessible. For example, to measure the cardiac output by the dye dilution method, a certain quantity of dye is introduced into the bloodstream and allowed to pass through the heart. The concentration, after it has gone through the heart, is produced as a curve. The cardiac output is obtained from the area under the curve rather than from the curve itself, and so an integration is necessary. The computer does this and presents information as cardiac output.

6. *Pattern Recognition*—In order to analyze bioelectric signals, it is

necessary to identify important features of the physiological waveform such as amplitudes and intervals. For example, computer programs are available to search the data representing an ECG signal for certain predetermined characteristics that identify types of heart problems.

8.4 THE FUTURE

What does the future hold? The bionic men and women depicted in popular television programs are indeed a possibility. We now have artificial arms that surpass the power of an ordinary arm, and some day we may have the versatility to make them practical.

One major question is that of artificial organs. In recent years, we have heard of many heart and kidney transplants. Will we be able to design and build artificial hearts and kidneys that we can implant? Both artificial hearts and kidneys have been demonstrated, but many problems remain. Many such units are much too large to fit in the body, and others that do fit are rejected by the body.

We have been fairly successful in the use of artificial parts. We have produced artificial heart valves, hip joints, aortas, and other arteries, to mention a few. The problem of body acceptance is a big one. One early experimenter replaced blood vessels with some made of nylon. The body simply ate the arteries up. They were just another protein. The whole problem of so-called "biomaterials" is in its infancy.

Some organs (for example, the heart) need power. There is the possibility of the use of nuclear power. Nuclear power has been used in pacemakers and in a model of the heart. However, what about radiation? Presumably the amount of power needed is so small, the radiation would be negligible.

Artificial limbs have been produced that are quite close to the real thing, and this field is well advanced. The major problem is to bring the devices down to a cost that people can afford. Likewise, assist devices for paraplegics have advanced considerably. The use of tongue switches for the limbless individual is a good example, and work continues. Tremendous advances have been made with artificial hip and knee joints. Also, a new field of rheumatoid mechanics is developing for helping people who lose the use of hands from rheumatism and arthritis.

Another item worthy of discussion is the use of body signals to activate dead or lost limbs. One illustration is the Boston Arm developed by

M.I.T. and Harvard Medical School at Massachusetts General Hospital a few years back, but similar principles are used extensively today.

The Boston Arm employs natural electric signals from unused muscles in the amputee's stump. In one trial, a fifty-five-year-old man was able to operate the arm with signals from stump muscles that had lain unused for twenty-six years. It took him only one fifteen-minute training period to restore control over those muscles. Another amputee learned to control bending and extending of the limb in five minutes.

The Boston Arm can be made with varying degrees of force, in the same way that a normal arm operates by "thinking" the amount of effort needed and translating it into the force required for a given task. When a person wants to move a normal arm, he begins by willing it into action. A cell from the brain sends a signal out to the motor nerve, which stimulates the muscle. The muscle shortens and the limb moves as desired. Although complicated, the operation works at a subconscious level.

The Boston Arm takes advantage of the fact that when a muscle contracts, it gives off an electric signal. If a person loses an arm from above the elbow, the muscle ordinarily is retained in the stump. If an amputee wills a nonexistent arm into action, the stump muscle will discharge the same electric signal as though he had an arm. With the Boston Arm, the electric signal is amplified and used to control a battery-powered electric motor in the arm. The battery is worn on a belt around the waist. The motor takes the ampified signal and powers the arm into action.

The use of signals for dead nerves is another place where much progress could be made in the future. Some ideas have been developed for deaf persons, but not too successfully. There have been demonstrations of television cameras to restore sight to blind people. These may sound like miracles, but they are feasible.

On the other hand, with all our knowledge, we are still very ignorant in many ways. An example is the measurement of blood pressure. Blood pressure varies all the time during each heartbeat. Since the heart's pumping cycle is divided into two major parts, systole (period of contraction) and diastole (period of dilation), blood pressure is usually measured by taking a maximum value (systolic) and a minimum value (diastolic). A typical value is 120 millimeters of mercury for systolic and 80 millimeters for diastolic, usually depicted thus: 120/80. The present method of measuring blood pressure was discovered eighty years ago and has not been improved upon. It involves using an inflatable cuff to occlude the blood supply temporarily and by use of a stethoscope and listening to sounds in the flowing blood making a determination which can vary from doctor to doctor and nurse to nurse on the same patients. In other words, a very crude measurement.

This is called an indirect method. We can measure blood pressure directly by inserting a tube into the bloodstream surgically, but this can only be done in a hospital and is traumatic to say the least, and a little dangerous.

The question the reader must ponder is, why have we not discovered a better method in eighty years? It is a paradox in a sense when we consider what other progress we have made in engineering applied to medicine. We must be able to measure this vital physiological parameter accurately and without danger or trauma. This is the type of problem that faces the biomedical engineer of the future.

What about body repairs? Many physiological adhesives have been developed. Complex organs like livers and kidneys have been glued together after combat wounds in Southeast Asia. Sprays have been developed to reduce pain, and recently a physician in New England has demonstrated a treatment to cure cancer which involves spraying a liquid on the skin that freezes the inside of an entire tumor and effectively kills the cancer. This has been effective in almost two thousand patients with a 97 percent cure rate.

This leads to another type of development in the use of low-temperature effects. That is cryosurgery, the use of freezing techniques during surgery. By lowering body temperatures, many things can be done that cannot be done at normal temperature. Also, this often gives the surgeon more time to perform the operation.

Another technique which may be way in the future is the use of *hyperbaric chambers* with which surgical operations would be done under an atmosphere a few times greater than normal. Since the amount of oxygen increases with higher air pressure, when a person is placed in a pressure chamber, the blood will absorb more oxygen, so that even if the heart is less than adequate, the person will not suffer from oxygen deprivation. Some work has been done in the USA, but in the USSR they are really keen on this technique. The Institute of Experimental Surgery in Moscow has a whole section of a hospital devoted to hyperbaric surgery. They feel one of the main uses is for obstetrics. If an expectant mother has a serious heart complaint, it is often necessary to terminate the pregnancy. However, if the baby is delivered in the pressure chamber, the greater oxygen saturation takes the load off the heart.

These units are very complex. Besides the physiological instrumentation necessary to monitor blood pressure, heart rate, etc., the atmosphere of the chambers needs to be monitored and controlled for temperature, pressure, humidity, gas composition, etc. Also, because of fire hazards, all measurement and control equipment has to be brought out external to the chamber, by means of wiring from the electrode terminals within the chamber. Some people have stated that they see a tremendous future in this technique. Others feel it is a waste of time, but who really knows until it is tried enough?

Most of this chapter has been concerned with diagnostics, repairs, and cures, but what is the biggest challenge in medicine today? Maybe it is to keep people healthy or to prevent their getting sick. Hence the term "preventive medicine." What is meant by this term?

Many people are very conscientious about servicing their car, but do they service their body? There is a growing school of thought that we should. Biomedical engineering becomes involved because of the fact that we have to monitor the body-variables in order to know how healthy we are and to detect signs of change. We do this by *medical screening*, sometimes referred to as *multiphasic screening*.

The ideal concept is medical checking from the cradle to the grave. This would be an extremely expensive program to mount. The problems are many and include:

1. *How often should you check?*—Should it vary with age? (For example, babies and older people would be checked much more frequently than persons aged twenty through forty. Every age group could be different.)
2. *What should be measured?*—Physicians disagree on what the most important data should be. Blood pressure, heart rate, blood chemistry are very important, but how far do you go to be effective?
3. *How complex should the facilities be?*—This must be considered from the viewpoint of physicians' and patients' time efficiency, capital outlay, maintenance, etc.
4. *How automated should this be for quick and accurate data?*—How much use should be made of computers?
5. *How should the data be studied?*—What are the diagnostics involved?

There are diverse centers today under many different names such as HMO (Health Maintenance Organization), HAC (Health Appraisal Clinic), and MHTC (Multiphasic Health Testing Center). Their objectives include: determining the fitness of an individual; developing health profiles, counseling on health matters including any necesary diet, exercise, or therapy needed; and establishing a good health profile for helping the individual plan the rest of his or her life.

Is this the way of the future? Can we afford such systems? Maybe the student can ponder this and decide, but there is no doubt that even if we have to be content with only a part of this system, biomedical engineering will play a vital role in its evolution.

8.5 THE CONSEQUENCES OF BIOMEDICAL ENGINEERING

To conclude this chapter it is necessary to take a look at the social and moral problems that might be involved and also at possible psychological effects. Some of these will be discussed.

Many persons are literally afraid of machines and devices. The fear may be because of their founded or unfounded dread of physical harm. Many physicians have impeded progress because they did not understand a diagnostic machine or the use of a computer. Nurses have been scared to touch controls.

With a patient this fear is amplified. Many people are terrified of needles and having a "shot" is a traumatic experience. Likewise blood tests can be terrifying. Many of these fears have substance. Persons have been electrocuted by electronic medical devices, but for every accident there are millions of safe results. Nevertheless, the apprehension of having, for example, an electrocardiogram taken is often great. Some tests with no danger whatsoever can have very strange effects. The taking of blood pressure by the use of a cuff, a method with virtually no danger, can scare a patient so much that the data obtained are not trustworthy.

Today biomedical engineering represents just a few percent of the multi-million-dollar health-care industry, but if it is to grow, many of the fears of physicians, nurses, and patients have to be allayed.

Another aspect is the dehumanizing effects of mass-produced medical analysis. Everything is done by machine, results come out by computer, the patient is merely a number.

What about artificial organs, transplants, and implants? The problems relating to the organ donor will continue to plague man as long as organ transplants are performed. When in fact can an organ be taken from a dying person? The legal problems are formidable. If a physician takes an organ from a dying person before he is legally dead, the doctor could be charged with murder, yet who is to say that a person is dead? If his heart is still beating is he alive? Today it is generally accepted that if brain function has ceased, or death is obvious, organ extraction can take place. This definition of death essentially leaves the choice up to the consulting physician. Therefore, if the doctor can give sufficient reason for his choice, he cannot be prosecuted. There is always room for human error, and the short time allowable between extraction and transplantation only increases the possibilities of errors. Problems concerning the donor's death are more common when the operation involves a heart or liver transplantation or other transplant where only one organ is common to the human body.

Every normal person has two kidneys and can live adequately with one. However, what happens if, after donating one of his kidneys, the donor's other kidney becomes diseased? His donation has put his life in jeopardy. Does he have the right to take back the kidney? What value is a kidney transplantation if it lengthens the recipient's life by a few years but shortens the donor's life? These are a few of the moral problems surrounding the question of donation of organs.

There is also the difficult question of who the donee should be. This question also applies to the selection of hemodialysis recipients. Of the approximately fifty thousand people who die of uremia every year, seven thousand are qualified for kidney transplants or hemodialysis. Only one thousand of these patients can be accommodated. How are these life-death questions resolved? Our religious teachings tell us that each person has a right to life no matter what his social background, but somehow these decisions must be made, sometimes by the mundane fact of how much the patient can afford.

With the advent of advanced technology in biomedical engineering, it is possible to extend life almost indefinitely when the heart, lungs, or kidneys give out. To what point should a patient in deep coma be allowed to live with the support of such machines if the hope of recovery is small? Can a conscious patient who is terminally ill pull the plug on his own life-support machines? Today the legal and religious establishments condone the abandonment of life-supporting machines if eventual death is certain, or if the psychological and economic burdens of continued life are extreme. The main problem with this guideline is that death is never one hundred percent certain. There have been many cases of complete recovery after deep coma and seemingly terminal illnesses. In addition to this, the rule does not clearly define how heavy these burdens must be. Clearly life-death decisions are filled with uncertainty.

Psychological effects on the recipient of a transplanted organ can be harmful. Often the recipient loses his will to go on. This results from the patient's loss of self-confidence and self-sufficiency. A similar condition usually develops with hemodialysis patients. States of depression also may result from the patient's concern over being a burden to his family. While the patient may become psychologically affected, there is a greater chance that the trauma of transplantation or hemodialysis may affect his family to a greater extent. Family members have been known to be troubled by the fact that their relative has an organ inside him that once belonged to somebody else, or conversely that a family organ is implanted in a stranger. These problems do not seem to be as intense when transplants take place in families, such as in twins where one has a deficient organ.

What about transplantations and implantations concerning the brain? When considering the possibilities, the prospects of behavior control surface. Many feel that tampering with the brain violates the highest moral standards of human dignity. Could not brain modification be allowed if it were to correct a socially disagreeable behavior or a physical disorder of the nervous system? Again, as with transplantations, guidelines would have to be set up to govern the actions of physicians.

To conclude, in the future we may even be able to create new forms of life by genetic engineering, but the social, moral, legal, and psychological issues posed are endless. The question is, "Where do you draw the line?" This is a conclusion every student must search for in his own mind.

8.6 EXERCISES

1. How would you define the field of biomedical engineering?

2. How would you think a clinical engineer differs from other types of bioengineers?

3. What would you consider to be the objectives of a biomedical instrumentation system?

4. What do you believe are the social consequences of the marriage of the fields of medicine and engineering?

5. Name six body functions and relate them to a field or topic normally studied as an engineering-type or physical subject—for example, cardiovascular system and fluid mechanics.

6. What do you understand by the term "man-instrument system"? Give an illustration and draw a block diagram of a system—for example, a medical technician taking an ECG of a patient.

7. What is the difference between diagnostics and therapeutics?

8. What is a biopotential? Give some examples of biopotentials.

9. What do you understand by the term "traumatic measurement"?

10. What is a heart-lung machine? Why is it useful for surgery? .

11. What do you understand by the term "pacemaker"?

12. As yet, nobody has defined a field of "psychoengineering." Would you care to define it?

13. What is biotelemetry?

14. Write a short essay on how you could see engineering principles and practices used in a hospital.

15. Discuss some of the uses of a computer in medical applications.

16. What do you think of the possibilities of an artificial heart powered by nuclear energy?

17. Comment on measurement of blood pressure.

18. What do you understand by the term "preventive medicine"? How do you think engineering can help in this field?

19. What do you think has been the greatest idea that engineering has applied to medicine? Give reasons for your answer.

20. Present some ideas of where you think engineering can help medicine in the future.

21. Write a short essay on your feelings about a kidney transplant (a) as a donor; (b) as a donee.

22. Discuss the fears of a patient when subjected to measurements to himself with electronic equipment.

23. Write a short essay on the merits of keeping a patient alive by means of life-sustaining machines.

24. What are your views on the creation of artificial life by means of genetic engineering or any other method that may appear in the future?

8.7 BIBLIOGRAPHY

1. Bekey, G. A., and M. Schwartz. *Hospital Information Systems.* New York: Marcel Dekker, 1972.
2. Berkley, C. "Medical Research Engineering—Past and Future." *Medical Research Engineering.* December 1971.
3. Brown, J. H. U., J. E. Jacobs, and L. Stark. *Biomedical Engineering.* Philadelphia: F. A. Davis, 1971.
4. Cromwell, L., M. Arditti, F. J. Weibell, E. A. Pfeiffer, B. Steele, and J. A. Labok. *Medical Instrumentation for Health Care.* Englewood Cliffs, N.J.: Prentice-Hall, 1976.
5. Cromwell, L., F. J. Weibell, E. A. Pfeiffer, and L. B. Usselman. *Biomedical Instrumentation and Measurements.* Englewood Cliffs, N.J.: Prentice-Hall, 1973.
6. Dickson, J. "Bringing Technology into Biomedicine." In *Dealing with Technological Change.* New York: Auerbach, 1971.
7. Fishlock, D. *Man Modified.* New York: Funk & Wagnalls, 1969.
8. Geddes, L. A., and L. E. Baker. *Principles of Applied Biomedical Instrumentation.* Second edition. New York: Wiley, 1973.
9. Hellman, H. *Biology in the World of the Future.* New York: Evans, 1971.
10. King-Hele, D. *The End of the Twentieth Century?* New York: St. Martin's, 1970.
11. Ray, C. D., ed. *Medical Engineering.* Chicago: Year Book Medical Publishers, 1974.
12. Rushmer, R., and L. Huntsman. "Biomedical Engineering." *Science.* February 1970.
13. Weiss, M. D. *Biomedical Instrumentation.* Philadelphia: Chilton, 1973.
14. Wertz, Richard W., ed. *Readings on Ethical and Social Issues in Biomedicine.* Englewood Cliffs, N.J.: Prentice-Hall, 1973.
15. Williams, P. N., ed. *Ethical Issues in Biology and Medicine.* Cambridge, Mass.: Schenkman, 1973.

Chapter Nine

GENETIC ENGINEERING: TINKERING WITH LIFE?

Genetic engineering is a very new branch of engineering which deals with genetic transformations selectively. In recent years, scientists and engineers have developed techniques for deliberately splicing into one organism's DNA pieces of genetic material from another. DNA, which is short for deoxyribonucleic acid, is the active substance of the genes of all living things. It therefore governs the heredity of all life on earth. The successes in selective gene-splicing, i.e., replacement of defective or undesirable genes by functional and desirable ones, will force mankind to face squarely many issues of ethics and values, some of which are discussed in this chapter.

First it seems helpful to separate genetic interventions used for *therapeutic* purposes from those used for *breeding* purposes (that is, to "order" a child with specified attributes). Next, it seems useful to distinguish between genetic interventions introduced to serve *individuals* (e.g., parents who desire a normal child or a child of high IQ) and those used to promote *societal,* or public, policy (e.g., eradicate disease, breed wiser people). We must also make a distinction concerning the *method* of intervention used,.dividing societal services into *voluntary* controls (e.g., asking, but not forcing, people to limit their family size) and *coercive* controls (e.g., laws requiring couples to take a Wasserman test to rule out syphilis before they marry).

9.1 INDIVIDUAL THERAPY

The first cell generated by the intersection of these coordinates is the easiest to deal with. There is little reason for not providing individuals with all the genetic therapeutic services they are willing to use. Surely no church or government should force parents to give birth to severely deformed children, and to force into the world children doomed to a distorted, miserable life. Genetic counseling, mass screening, and amniocentesis should be available to all. Many of these genetic interventions, like many other forms of advanced medical services today, mainly help the well-off, largely because the poor are less informed, more economically constrained, and less likely to seek medical assistance. This is one of the tragedies of our society.

Even for individual therapeutic goals—their most obvious and beneficial application—new genetic techniques should be made available only after they are well tested. In the field of genetic intervention, there are fashions and fads and occasions when new devices are made available to people before they are effective or safe. A well-known example: a PKU test* was rushed into use before it had been perfected, as a consequence of which previously healthy children were put on rather damaging diets. New genetic procedures should be examined by a powerful review board, like the kind that now reviews drugs before they are marketed. But such a Health-Ethics Commission would have to be more effective and potent than the present FDA; it would have to curb both corporations seeking to make a quick buck from new techniques, and political and medical headline seekers—the health hot-rodders like those who rushed the PKU test through.

Once genetic techniques were proven sound, we might curb their application only under one condition: if there were strong evidence that providing such services to individuals would cause serious and present harm to *society*. Service to individuals might be limited, for example, if studies showed that by providing amniocentesis on demand the genetic foundation of the human race would indeed be seriously endangered—not on some future day, when all of civilization may break down, but here and now.

The individual cannot be granted an unlimited priority over society, if only because the individual is part of society and needs it for his or her

---

*A test to determine retardation of an unborn child caused by phenylketonuria.

survival and well-being. Therefore to curb pollution in downtown Los Angeles and in some New York City tunnels, where pollution endangers the welfare of local residents, we are quite correct in limiting the use of autos and urging people to drive in auto pools or take buses or trains. Even in this circumstance, when society's need is urgent, it is often more practical and more ethical not to force people to change their preferences (here, for cars) but to try to use new technologies to reduce the societal cost (e.g., to render cars less polluting). Similarly, if some new genetic interventions cause problems, we should, before we ban them, see if their side- or after-effects can be eliminated. But if such measures fail, or until they are available, society is entitled to limit services to individuals.

In genetics such curbs are especially difficult to justify (and hence should be introduced less frequently than in most areas) because we are dealing not with a convenience or even an economic need, but with a very intimate, personal part of our lives. It is one thing to forbid people to drive their automobiles downtown or to require the use of seat belts; it is quite another to force them to be sterilized, to prevent them from obtaining safe abortions, or to force a woman to bear a mongoloid child. As such the test for curbing genetic services must be particularly exacting. Fortunately, in most genetic matters either individual desires seem to complement societal ones, or services to individuals, such as genetic counseling and abortion, have only minor effects on the societal gene pool and therefore on the quality of the human stock.

9.2 SOCIETAL THERAPY

But what about a situation in which the *society* has therapeutic goals but the individuals don't go along with them? At first glance, this may seem absurd; how can we speak about a society that has therapeutic or any goals other than those of its individual members? Whose goals are these? And why won't the individuals accept services aimed at improving their health or that of their unborn children? That this is quite possible can clearly be gleaned from nongenetic areas. Take tooth decay and smoking, for example. Both profluoridation and antismoking campaigns are promoted by government. If individuals were really willing to heed sound medical advice on basic matters, these campaigns would be unnecessary. But because of irrational fears (in the case of fluoridation) or addiction (to cigarettes), society *does* enter the picture. In the case of smoking, society employs *voluntary* methods (propaganda) and economic pressure (high taxes on cigarettes). In the case of fluoridation, society uses *coercive* measures: fluoride is injected into the water mains in many communities without citizen consent or active knowledge precisely because, when the issue is put to a vote, it is often vetoed. Thus, society's forcing of its members to attend to their health is far from an unknown phenomenon. Could society step in, on the same grounds, concerning genetic matters?

Coercive genetics—the use of society's laws, courts, jails, and policemen—to weed out undesirable genes by force, seems intolerable and repugnant. Unlike excessive use of autos, or even the abuse of one's teeth, the union between two persons that gives life to a third should be kept free of all government intervention. It would be horrifying if the government budgeted the number of children per family or put a contraceptive drug into the water supply or forced mothers to abort "surplus children." One need only think of what would happen if some official decided that in order to reduce criminality, chromosome tests on all pregnant women would be required and abortions demanded of all mothers carrying XYY "criminal" fetuses. We would end up with policemen dragging women to abortion clinics and mothers going underground to protect their embryos. Government use of force with respect to these matters would constitute the ultimate violation of the contract that makes the state tolerable to the people; it would completely undermine the legitimacy and the moral basis of government.

Genetic technology will improve and the appetite to interfere may well be excited. If any attempt is made to move in the direction of coercive intervention, therefore, it should meet with the utmost opposition of citizens and their representatives. To symbolize and ingrain the rejection of forced genetics, all existing genetics-by-legislation, that is, by force of law, should be repealed. This includes laws that forbid marriages among the feeble-minded (such as those still on the books prohibiting marriage between men of any age and women under forty-five, if they have a history of insanity, are feeble-minded, or are imbeciles, habitual criminals, or common drunkards) and those requiring sterilization of the mentally retarded.

The way society actually feels about these matters is well reflected in the fact that these laws are almost never enforced. Having them on the books did not matter much when they did not set a precedent for other types of interventions, because there was not much that citizens could be coerced into doing in these matters. But as the possibilities of coercive interventions are now rapidly multiplying, wisdom calls for establishment of the strongest possible barriers against them.

The opposition here is to forced genetic interventions, not the setting and promotion of genetic goals. As individuals we may be shortsighted or selfish, acting as if we are the only ones in need and disregarding the fact that what might work for one often will not work if all individuals act in the same manner. Hence there is good reason to take into account societal needs—those future, aggregate and shared needs of the people who make up a given society. But these needs should be met voluntarily, not through the use of force. Which of the variety of voluntary means should be relied upon depends upon the circumstances. If the societal need is very urgent, if the burden is overwhelming, economic means can be utilized. For example, a government under great economic pressure, as in a severe recession where there is a great demand for societal services, or during a prolonged health crisis, might inform its citizens that the public institutions will no longer allow parents to dump deformed children on them. Thus, while no one should force parents to abort a mongoloid fetus—now that parents have a choice—society does not, under all economic conditions, have to pick up the tab for the upbringing of such children.

When societal needs are less pressing, persuasion without economic sanction, such as the kind used to get people to accept birth control and to curb drinking, should be chiefly employed. One can imagine ads saying, Give Your Child a Chance to Live a Full Life—Check Your Genes, or, You No Longer Must Bring a Mongoloid into the World. Other educational means, even organized tours for prospective parents through wards of deformed children, could be employed.

Public health officers may not wish to rush out wearing slogans like Stamp Out Genetic Illness because genetic illnesses cannot be overcome the way other diseases can be. Testing everyone's chromosomes and pulling out all the sick genes will not rid us of genetic illnesses. The basic reason is that nature continues to produce new supplies of such genes through mutations. These are not unlike printing errors; however carefully you set up the type, for every x thousand print-offs (or children conceived) a certain number will be defective. Thus even those genetic illnesses that result in the early deaths of all carriers *before* they have children do not disappear.

It has been suggested that the whole attempt at genetic public policy is a hopeless waste of time. But this is not the case. We may not be able to reduce the defective rate, but we can at least catch nature's errors and eliminate them before they produce miserable children, agonizing parents, and public charges. We may well have to repeat the process for each generation, but this does *not* render it valueless. While it would be ideal to eliminate these illnesses once and for all, the next best thing is to eliminate their consequences in human and economic costs.

The basic rights of an individual in a free society include that of having as many of whatever kind of children a person is willing to have; society can try to persuade people to have fewer children or to abort severely deformed ones, but it must not force these choices. However, the individuals' rights do not include the liberty to charge the upbringing of their children to the public. A society might go so far as to inform all prospective mothers, especially those in high-risk categories, that a genetic test is highly advisable, and further, to inform those whose tests show them to be carrying a deformed fetus that *they* will have to provide for it. But again, the use of genetic police or inspectors must never be tolerated.

9.3 SOCIETAL BREEDING

The use of genetic techniques for improving genetic qualities raises a quite different set of questions. This form of genetic intervention is discussed most often from a societal viewpoint because it is here that the best-known attempts at breeding "better" people have been made. In the past these efforts were directed toward goals that almost all people find abhorrent, notably those by the Nazis, who tried to breed a "master race" using such abusive techniques as the extermination of those whose genetic qualities they deemed "inferior" (not just Jews, but also feeble-minded Aryans and other populations). Within the German population, the regime imposed the compulsory sterilization of manic-depressives, severe alcoholics, the feeble-minded, epileptics, and those suffering from hereditary blindness and deafness, and the castration of dangerous and habitual criminals. To preserve the "hereditary soundness of the German race," marriage was forbidden when one of the parties had a dangerous contagious disease or suffered from mental derangement or a hereditary disease. Racial intermixing and intermarriage of Germans with foreigners were prohibited, and a German who did so could lose his status as a German.

The taboo we now have against the deliberate breeding of certain types of people is so effective as to bring to mind the label "racist"—

used so often to refer to people opposed, openly or covertly, to equal rights for minority members, especially to those who base their position on inherent, genetic differences between the majority and the minorities. Aside from reminding one of the Nazis, the very notion of selective breeding brings to mind the Ku Klux Klan. But is it time to examine these taboos? Can one simply dismiss out of hand all the "promised lands" that distinguished scholars point out are within our reach? Mankind is in sore enough condition; it seems rash simply to brand "unthinkable" the promise of a race characterized by "freedom from gross physical and mental defects, sound health, high intelligence, general adaptability, integrity of character and nobility of spirit" (4, p. 51). Is it not presumptuous to dismiss out of hand the notion of using biotechnology to create people with "a genuine warmth of fellow feeling and cooperative disposition, a depth and breadth of intellectual capacity, moral courage and integrity, an appreciation of nature and art, and an aptness of expression and communication" (8, p. 35)?

Nor can one easily dismiss the argument that the pendulum of public policy has swung much too far in the educationalist and revisionist direction, away from biological considerations. In typical dialectic fashion, we have moved from the thesis, popular in the first decade of this century, that man is governed by biological instincts (sex, hunger, aggression) to the antithesis of the educationalist concept of man. In 1925 John B. Watson, a founder of behaviorism, issued his famous challenge:

> Give me a dozen healthy infants, well-formed, and my own specified world to bring them up in, and I'll guarantee to take any one at random and train him to become any type of specialist I might select—doctor, lawyer, artist, merchant-chief, and yes, even beggarman and thief, regardless of his talents, penchants, tendencies, abilities, vocations, and race of his ancestors. (13, p. 104)

Other educationalists followed, especially once the instinct theory became associated with fascism in Europe and racism in the United States. By the late sixties, numerous programs, from labor training to compensatory education of the disadvantaged, from mental health clinics to cures for smokers, all assumed that education could readily improve the lot of anybody.

When it became increasingly evident that this assumption was not always valid, the interest in physiological and genetic factors was reawakened. The Coleman Report, published in 1966, probably marked the turning point. It raised difficult questions about the potency of education. Then came Arthur Jensen's and Richard Herrnstein's articles, which argued that IQ differences between blacks and whites were, to a significant extent, genetically inherited. Each of these documents had a much larger impact than a typical scholarly work, because they were publicized, discussed, and debated, and because the era was ripe for a reaction to educationalism.

The new era is unlikely to return us to Fascist notions of genetic determinism, but instead will move public policy toward a synthesis that would rely on both educational and biological factors. The synthesis era will be concerned with their combinations and interaction. Compared to the educationalist period now ending, the new era seems to show more interest in and tolerance for genetic engineering than ever before, but without going overboard and seeing it as a cure-all.

Finally, one cannot simply dismiss the notion that our drive to govern our condition, rather than remaining subject to the blind fluctuations of forces we can neither understand nor control, might be helped through biological engineering in addition to institutional reforms and power redistribution. The curse of modernity is that the revolutionary expansion of means—of instruments—has wreaked rebellion against the creator and his purposes. Like a Frankenstein monster, technology has gone beyond the control of its maker; it distorts society to suit the logic of instruments rather than to serve the genuine needs of its members. The primary mission for the next era is the restoration of the primacy of human values. This may be reflected, for instance, in the willingness to trade off at least some economic growth and technical progress for more humane work and a greater care of nature—in short, there may be a less competitive society.

The trouble is that, at present, all efforts to restore the primacy of human values over tools—by expanding our brain power and wisdom—have not progressed very far. Efforts to do so via institutional reforms, social revolutions, or rejuvenation of the self seem to provide at best only partial solutions. Hence one has to consider the notion, advanced by Glass, Muller, and others, that a "higher," less aggressive, more intelligent breed may have to be biologically cultivated before a more humane society can arise. Genetic engineering, one biologist has suggested, could help remedy such problematic societal proclivities as nationalism, aggression, and excessive bureaucratic inclinations (3, p. 7).

It is naive to believe that in large organizations, such as federal bureaucracies, the tendency to malfunction has a genetic base. But other attributes, such as aggressiveness, intelligence, level of energy—and hence achievement and motivation—are not these *in part* genetically affected? True, a person who is aggressive could be trained, that is, educated, to be a prosecuting attorney, a soldier, or an assassin, a point stressed by the educationalists. But is it not also true that, given an aggressive race, peace is going to be difficult to achieve and sustain whatever the educational reform efforts? And if people tend to be lethargic, can education turn them into a productive and creative people? It seems that we must draw on both societal and biological factors if the human condition is to be bettered. This conclusion may be obvious to some, but it is hardly so to those brought up in the mainstream of the libertarian or social-science traditions of the last generation.

Is such thinking "racist"? Only if one assumes that some groups of people have "bad" genes while others have superior ones. But racism is not at issue, for *all* of mankind's genetic stock may well stand in need of improvement: no one group or race has a monopoly on good or bad genes. Human breeders will be like those who, seeking a superior breed of cattle, try to combine the superior qualities of several existing breeds; the resulting hybrid will have little resemblance to any of the original races.

Just as no group has all the desired genes, so each subpopulation seems by nature to be afflicted with one or more undesirable genes. Blacks are afflicted with sickle-cell anemia; Eastern and Central European Jews with Tay-Sachs disease; Mediterranean stock, Cooley disease; Caucasians, cystic fibrosis; and so on. And even if there is a group that is worse off, this calls for more compensation and public service, not racial slurs.

Yet many have argued that such a breeding policy is possible for racehorses, hogs, and dogs—but not for human beings. We could never agree on what to breed: athletes, eggheads, redheads or blondes. The very attempt to do so would break society up in conflict.

But would not most of us want to breed, say, more intelligent and compassionate persons? And who said that we need a uniform race? Could we not breed some of each kind? Above all, since the implementation is to be *voluntary,* there will be no more uniformity than people choose to have. And no need to reach a consensus.

At issue is a public policy that welcomes certain biological features over others—e.g., energetic over lethargic qualities. As with our call for limiting family size, some are influenced by it, others ignore it. Similarly, in the case of breeding policies, even if there were one recommended fruit, many would dislike the taste and not buy it. And if some attributes do prevail—as we do now breed taller people because of widespread acceptance of the recommended use of vitamins—it would be only because many people accept the policy. There is as such no inherent contradiction between a genetic policy and a democratic society.

Several biologists have argued that this is all pie in the sky; that such breeding is technically infeasible in the near or even remote future. The whole argument is unnecessary, they say, because no such changes can be effected. First of all, as mandatory and uniform policies are out of the question, voluntary adherence would be limited in scope and hence in effect; second, to have the desired effect, some rather specific strictures would have to be adhered to. The example used concerns intelligence (an unfortunate choice, because intelligence is affected by several genes, and the effect is surely more complex and difficult to bring about than, say, changes in height).

These skeptics point out that in order to achieve the desired effect, women with low IQ's would have to marry men with high IQ's, or men with low IQ's women with high IQ's (the mating of two high IQ's does not make for a higher IQ), which is unlikely to be carried out voluntarily on a large scale; and—to repeat—coercion on behalf of such a goal is unthinkable and repellent. As to the use of artificial insemination, if ten percent of the women of one generation used the sperm of 160-IQ donors, the average IQ rise would still be a very low 1.5 points (6, pp. 96–99).

Other equally distinguished scientists are much more optimistic as to what could be done through genetic engineering. It has also been pointed out that while the average increase in IQ that might result from genetic engineering would be low, a mere increase of even one percent in the average would result in 3.5 to 4 million additional very high IQ (175) persons (6, pp. 102–4).

The truth would seem to lie, as it so often does, somewhere between the extremes of advocacy and derision. "Coercive eugenics," those imposed by the state, must be abhorrent to all, and must be fought by all means known to human beings, like other totalitarian policies. But there are also "voluntary eugenics," publicly promoted, freely accepted or refused, like participation in the March of Dimes. One ought not to confuse the two. Unless someone could bring up a new argument against a public policy that would encourage people voluntarily to favor certain traits—say intelligence or warmth—and if genetic promotion of these traits were technically feasible, then surely a limited genetic experiment (*not* a federal crash-project investment of five billion dollars to breed a brighter or more peaceful race) might be acceptable. It has been argued that even such a limited experiment would require twenty generations to complete because of the slow accumulative effect of such changes. The scope of the effect depends, of course, on how many people choose to heed the genetic suggestions and the extent to which they marry each other rather than "outsiders." If many participate and marry among themselves, the effect will be greater. And while the benefits will almost surely be gradual and not sensational, that is hardly a reason for opposition.

9.4 INDIVIDUAL BREEDING

One may still be reluctant to favor a public policy on the side of genetic improvements—but how about individuals shaping the genes of their offspring to their own hearts' desire? "Gene shopping," that is, choosing and combining the biological qualities of a child yet to be conceived and designing it to the parents' preferences, is discussed chiefly in science fiction and occasionally in the popular press. Because the technical means for gene shopping are not available now and will not be in the near future, scientists rarely view these questions seriously. To most experts in the field, the day when a parent can go to a "gene-mart" and ask the clerk for the appropriate genes to produce a blond, blue-eyed, tall, slender, high-IQ boy is so remote that they feel it is quite unnecessary to worry about the wisdom, ethicality, or social consequences of developing "genetic supermarkets."

Experts maintain that gene shopping is a long way off because most of those biological attributes that are genetically determined are controlled not by one, but by several genes, acting together in ways far from fully understood. Thus it might be relatively easy to shop one day for height or hair color (attributes relatively simply determined), but it is quite unclear what would have to be bought if one were seeking a high IQ, or many other desirable attributes.

Secondly, most attributes are affected by both the genes in the husband's sperm *and* in the wife's ovum. Hence, unless the trait is dominant you cannot shop for it without being willing to gamble. Some writers talk about buying the egg, too, and implanting it after it has been fertilized with the desired sperm, or about frozen embryos made to specification. But technically, such possibilities are even more remote than gene-shopping. Moreover, since these techniques require not just the use of easily obtained male genes, but also surgical extraction of ova and surgical replantation and fertilization in a laboratory, the procedure raises many more ethical issues than gene-shopping.

Thirdly, most of our attributes are shaped in an interaction between our genetic inheritance and our psychic and social upbringing. The shoppers who ask for high-IQ genes may be quite disappointed to find they have a clever but unmotivated, or smart-alecky, child, or one who misapplies his or her talents, or looks down on his "dumb" parents.

Last but not least, we know from breeding domesticated animals that when we push one attribute we tend to weaken others, ending up with a highly vulnerable, unbalanced species. Pushing a particular trait is achieved through pure breeding of a type, which is done through inbreeding; inbreeding inadvertently intensifies other genetic traits which may be genetic deficiencies.

These technical difficulties will have to be faced by all breeders, whether they breed to advance a public policy or fulfill parental desires. However, they will particularly limit what an individual can achieve. A society can benefit from aggregate, "statistical" benefits; for example, if its efforts lead to an average improvement, albeit not manifest in each person, and even if highly diluted by other, nonbiological factors, it still may benefit. But to parents interested in specific attributes for their next child, statistical changes over a whole generation of children are of little interest; they want *their* child to be brighter, taller, or whatever, and individual change is particularly difficult to achieve. All in all, then, gene shopping seems to hold much less promise than the press or its outspoken advocates have claimed.

It is useful to discuss the implications of breeding to individual order, however, because *some* gene shopping is technically possible right now. While the available procedures are primitive and very costly, the issues they raise, at least psychologically, are not different in principle from those raised by "future" developments. Indeed, precisely because of early technical and related social developments (social acceptance or disapproval often affects later innovations), the question must be faced, and the sooner the better.

Sex-choice is a case at hand. The *same procedure now available to control mongolism*—the combination of amniocentesis with abortion—can today secure an infant of the desired sex. When amniocentesis is performed to determine if the fetus is mongoloid, fetal sex can be determined as well. Many doctors will not inform the parents of this finding because they do not wish the sex of the fetus to be used in abortion consideration. So far, there seems to be no case on record in which a doctor agreed to abort a fetus because it was not of the desired sex. (One doctor reports that he was tricked into abetting such an action by parents who asked for amniocentesis to check against mongolism; told they had a normal female fetus, the parents proceeded to arrange for an abortion, for they wanted a boy [1, p. 408].) But just as a doctor can be found for any other illicit purpose, it is just a matter of time before this is done—if it is not being done already, off the record. Also, the development of sex-choice techniques may make it possible in the *near future* to choose sex without abortion, by separating male- from female-producing sperm (7, p. 784; 10, p. 329; 5, pp. 87–94) (followed by artificial insemination), or by providing the woman with a douche inhospitable to one of the two kinds of sperm, which would make sex-choice as easy as applying a medication (12, p. 304). Success is more likely here than with most traits, because the mother's ovum plays no active role in determining the outcome. Thus sex choosers have to deal with only one ingredient, and that the more manipulable of the two.

The question, then, cannot be avoided. Should parents be allowed to choose their child's sex, and, by implication, other genetic qualities—a choice which obviously is not one of health over illness, not a therapeutic matter at all, but clearly a matter of breeding—the way a person may choose a doberman over a poodle or the other way around?

As long as sex choice entails abortion, one may say that, because of the marginal risk to the mother (and her future children) that is involved, this procedure is tolerable solely for therapeutic purposes. Yet we already allow—indeed, at least indirectly encourage through zero-population-growth propaganda—parents to choose abortion in order to limit their family size; hardly a therapeutic goal. We say that both parents and children would be happier with no more than two children to a family. Now should doctors or the state be empowered to decide that parents may not plan their individual children—their sex and, eventually, other attributes—but only their number? And what if a family of four boys feels one girl is essential to its happiness? Should not that decision be left to the parents?

The aggregate consequences for society of its members freely choosing the sex of their children would indeed be undesirable, but probably not severe enough to warrant limiting the development of sex-choice techniques, even if we could curb them or prohibit their application for breeding purposes. Recognizing that society has needs of its own and that a severe sex imbalance could damage these, we calculated a projected temporary sex-ration imbalance of up to seven percent a year male surplus (probably less following the impact of women's liberation) (9, pp. 55–56). Not a very serious damage.

Since society is not likely to be seriously undermined by such techniques, we should not prevent individuals from gaining whatever happiness they can. If this entails adding a boy or a girl to their family, why not let them?

The same would seem to hold true for another means of genetic shopping now available. Artificial insemination (AID) is used now to help couples who cannot conceive because the husband is infertile. Although as a rule no information about the donor is given to the parents, doctors tend to try to choose one whose characteristics closely resemble those of the parents. Most doctors would refuse to provide AID where the husband is fertile or allow the prospective parents to choose among donors according to some desired attributes; say, short parents requesting a tall donor to obtain a tall child.

Here the risk is not medical but psychic. In some cases fathers later resent children born as a result of artificial insemination, and mothers have been known to become infatuated with the unknown biological father. These cases are used by some to support the thesis that artificial insemination should be used only when all else has failed, and not for the biological designing of children. But firstly, data show that these tensions can be handled by carefully explaining the issues to the couples involved and by providing psychological counseling. In any event, the cases of serious emotional trouble are very rare (2). Secondly, the individuals' happiness and their preferences should prevail. Doctors may well alert parents to the psychic dangers (as they do when the husband is infertile), but beyond that, it should be up to the prospective parents to decide. (Doctors who feel AID violates their personal ethics should not, of course, be forced to provide it any more than a Catholic doctor should be made to perform an abortion. If the health authorities and medical societies legitimate the service, prospective parents should have little trouble finding willing doctors.) Thus, if a couple who are short in stature feel very strongly that they don't want to impose that condition on their child, why not allow them to get sperm from a tall, anonymous donor?

Initially parents may expect too much. Many attributes are not inherited or are inherited only in part, and each sperm contains a large variety of genes, differing in combination from the same father's other sperm. A mother with an IQ of 100, receiving sperm from a 140-IQ donor, cannot be sure that her child will have an IQ of 140, or 120, or even higher than 100. Moreover, as long as gene-by-gene shopping is not possible, the mother must "buy" *sperm*—the whole package of genes, which may include attributes she does not seek. To offer parents the "packages" they desire, sizable sperm banks would have to be established. Nevertheless, some attributes can be ordered, and for others, one can take a gamble, which is after all what we do when we conceive naturally. We allow people to gamble on winning a fortune in a state lottery; why prohibit them from trying to improve the biological lot of their children?

The next step may well be that sperm banks (which already exist; they have been set up for fathers who wish to preserve some of their sperm when they undergo vasectomy or are exposed to radiation in their work [2, p. 32]) will be used to store sperm from anonymous donors, typed according to their attributes. Parents or unmarried women would prepare a list of specifications for their next child and give this list to a sperm teller at the sperm bank who would check the files to see if all the specifications or only certain combinations of specifications were available. A fee would be paid and a vial issued. Sperm might one day be available in the form of a suppository that could be used without a doctor's assistance. Individual breeding would then be on its way and might even become fashionable.

Two matters should concern us if we proceed in this direction. The first is that prospective users be well informed as to how much can be achieved through AID. If sperm shopping catches on, it surely will necessitate a program of consumer education to discourage excessive expectations and tensions which might result. The various consumer watchdogs should make sure that the sperm banks don't engage in false advertising, oversell their product, or mislead prospective parents. And if corporations get into the genetic picture they will have to be stringently regulated.

Second, one must consider the notion that these genetic interventions, which rely on artificial insemination, are a form of adultery and immoral union. Theologians suggest that AID, especially when the husband is fertile, will further undermine the family. Many religions see it this way, and so do large segments of the public.

Society does not see fit to enforce the law forbidding adultery. When both parents consent, and the procedure is conducted in the cold medical sterility of a doctor's office, such a union has little to do with assignations.

Genetic intervention for individual therapy needs more support, not less. Societal force should not be applied for either health or breeding purposes. Voluntary promotion of public policy in genetic matters, for either of the twin goals, makes good sense, though not much can be gained for society because of technical limitations. Individuals should be free to breed as they will. Steps, however, must be taken to see that the public is better informed as to what to expect, so that they can make wiser decisions.

9.5 EXERCISES

1. Write a short essay on the National Institute of Health's guidelines permitting some recombinant DNA experiments while continuing the ban on others.

2. To what extent would you advocate coercive intervention on genetic matters? Do you know any present examples of intervention?

3. Do you agree that every individual in a free society has the right to have as many children (whatever kind they might be) as he or she wishes? Discuss.

4. Argue John B. Watson's famous statement on behaviorism quoted in this chapter. Refer to reference 13.

5. What problems of society could be solved by genetic engineering?

6. Man always considered, and often succeeded in, achieving a superior breed in plants and animals. Should he be allowed to experiment on himself as well? Discuss.

7. Cite some practiced breeding cases that are without the "engineered" genetic interventions.

8. Write a short essay on man's moral responsibility to the unborn child.

9. Do you foresee the possibility of "gene shopping"? Discuss this concept and support your ideas with literary evidence.

10. Discuss whether the parents should be allowed to choose their child's sex.

11. Write a short essay on artificial insemination.

9.6 BIBLIOGRAPHY

1. "An Abuse of Prenatal Diagnosis," Letter to the Editor, *Journal of the American Medical Association.* vol. 221, no. 4, July 24, 1972.
2. Behrman, S. J. "Artificial Insemination." In S. J. Behrman and Robert W. Kistner, eds., *Progress in Infertility.* Boston: Little, Brown, 1967.
3. Danielli, James F. "Industry, Society, and Genetic Engineering," *Hastings Center Report* vol. 2, no. 6, December 1972.
4. Glass, Bentley. *Science and Liberal Education.* Baton Rouge: Louisiana State University Press, 1959.
5. Gordon, M. J. In *Scientific American.* vol. 199, 1958.
6. Leach, Gerald. *The Biocrats.* Revised edition. Baltimore: Penguin, 1972.
7. Lindahl, P. E. In *Nature.* vol. 181, 1959.
8. Muller, H. J., in H. Hoagland and R. W. Burhoe, eds., *Evolution and Man's Progress.* New York: Columbia University Press, 1962.
9. Peel, John. "The Hull Family Survey." *Journal of Biosociological Science.* vol. 2, 1970.
10. Schröder, V. N., and N. K. Koltsov. In *Nature.* vol. 131, 1933.
11. "Sperm Banks Multiply as Vasectomies Gain Popularity." *Science.* vol. 176, April 7, 1972.
12. Unterberger, F. *Deutsche Medizinische Wochenschrift.* vol. 56, 1931.
13. Watson, John B. *Behaviorism.* Revised edition. Chicago: University of Chicago Press, 1962.

On the morning of September 25, 1978, Pacific Southwest Airlines flight 182 was struck by a small plane over the city of San Diego. All 135 persons aboard were killed instantly when the plane crashed into the city. Dr. Alan Tetelman was one of those passengers. He was en route to an air safety conference being held in San Diego that day.

Chapter Ten

PUBLIC RISK AND ENGINEERING SAFETY: HOW SAFE IS SAFE ENOUGH?

10.1 INTRODUCTION

This chapter deals with the meaning of risk and safety and the various methods that society utilizes to assure that there is a reasonable balance between them. Society, in the form of governments, corporations, and individuals, builds and operates engineering systems to meet human needs. An engineering system is a process or device in which matter, energy, or information in one observable and definable condition is transformed into another definable condition, or is maintained in its original state when subjected to external forces. The total cost (in dollars) that society, in the aggregate or as individuals, spends on a given system is a measure of the worth or *benefit* of that system. The total cost is primarily that required to design, manufacture, operate, or maintain the system.

All systems are subject to possible malfunction and/or failure. Additional costs must therefore be allotted to system replacement (e.g., the value of an aircraft that has crashed) and to other systems and people who are damaged as a consequence of the failure (e.g., widows and children of wage earners who die in a plane crash). The *hazard* or *risk* that is associated with a given system is the cost (dollar effect) of failure. Since the effect of a failure can range from a minor incident to a major catastrophe, an entire spectrum of consequences can be associated with a given failure event. However, to simplify discussion, it is often convenient and informative to speak of the average consequence or average severity that results from a failure event.

Systems are designed to run properly, without failing, and this requirement is usually met. Those few failures that do occur in the life of an operational system are isolated instances, chance happenings, that are called "accidents." An accident is defined as "one or a combination of unanticipated failure events (incidents) that release stored energy, which directly or indirectly causes personal injury, property damage, or malfunction of a system." An accident is an event that is decribed in terms of both a "cause" and an "effect."

Despite the tremendous increase in the safety awareness of our society in the last decade, and despite the various methods and technologies that are available to prevent accidents or reduce hazards, failures and accidents can never be completely eliminated from our society. The real questions society must address are not "How do we guarantee zero defects?" or "How do we achieve complete technical security and freedom from failure?" but rather "How do we minimize the rate of failure?," "How do we reduce the consequences of a failure?," and "What level of total risk is acceptable to our society?"

When a consumer purchases a household product, or a military service purchases a new weapons system, a conscious decision has been made that the cost (to the purchaser) is less than or equal to the actual and perceived benefits that are associated with the product. Until recently, the cost of failure and accidents was crudely accounted for in the total cost on the basis of past performance of the product. The number of expected failures would be estimated and multiplied by the average cost of a failure. This amount would be set aside (e.g., in the form of insurance premiums) to be used to cover the cost of repair and replacement and recompensation after an accident took place. Since failure and accidents were rare events and the cost of failure was minimal, little societal attention was given to questions of safety and acceptable risk.

In the last decade, society's interest in and its approach to safety have changed greatly. A full discussion of the reasons for the change are beyond the scope of this chapter, but several major factors should be noted. First, with the development of larger systems, the magnitude of a catastrophe such as an airplane crash has greatly increased. Improved communications, primarily through the evening news on television, has brought these catastrophes close to a large body of the public, in a dramatic fashion that affects individual senses. The development of nuclear weapons and nuclear power has led to public concern for long-term, genetic effects of accidents (e.g., due to leaks of radioactive waste), that cannot be quantified, nor accounted for in terms of insurance premiums paid by the manufacturer of the product. Finally, there is now a public demand for increased corporate concern for the welfare of the individual and his natural environment. This demand was originally associated with the efforts of Ralph Nader, but today a host of environmental-action and public-interest groups closely scrutinize governmental decisions that directly and indirectly affect public safety.

In developing answers to the question as to what constitutes an acceptable or reasonable risk, several factors must be kept in mind. First, that acceptable risk to one group may be unacceptable to another. Second, that there is often a vast difference between the actual, quantified level or risk and the level of risk that is "perceived" in the minds of individuals or groups of individuals. Third, as stated earlier, that zero-risk is an unreasonable, utopian ideal. Finally, that lower risk can be achieved by reducing benefits (e.g., there is no risk of air crashes if no aircraft are flying) or by increasing the cost of safety systems.

The purpose here is not to provide definitive solutions to risk problems, but rather to explain basic concepts of accidents and risk. With an understanding of some basic concepts, the reader should then be able to explore the more complex questions in those areas of direct interest to him. Since risk is directly related to the frequency and severity of accidents, it is appropriate to begin with a description of the basic elements of an accident in a technological system. A following section describes the means by which the risk of accidents can be quantified, so that there exists a method for comparing the safety of one system with that of another. The next section describes some parameters by which society has determined levels of acceptable risk, while the final section describes the means by which society maintains a balance between risk, cost, and benefit.

10.2 ACCIDENTS IN TECHNOLOGICAL SYSTEMS

The conversion of matter and/or energy from one form to another in an engineering system is best described by a simple block diagram containing inputs and outputs. The inputs are combined at the transformation point (also called reaction point or node) into one or more outputs. The input/output process is observed and monitored by a controller (governor, regulator) which monitors the rate, direction, and magnitude of the inputs and outputs. Let us consider, for example, a steam-turbine-powered pump that operates in a refinery (Figure 10.1). Hydrocarbon liquid at pressure P_1 enters the pump, which is driven by a steam-turbine shaft. The turbine shaft is driven by steam at a rate that is controlled by a governor. The turbine shaft turns the compressor wheel in the pump such that the liquid exits at a higher pressure P_2. The governor monitors the speed of the turbine shaft by operating a valve, such that if the turbine slows too much, steam enters the turbine and increases its speed and, conversely, less steam is allowed to enter the turbine if the turbine starts to overspeed. A second governor is often added to the system to prevent gross overspeeding. In this subsystem of a refinery, the compressor, turbine, and governor are called *components;* the shafts, bearings, casings, and valves are principal *parts* of the pump and turbine.

FIGURE 10.1

SIMPLE PUMP/TURBINE SYSTEM

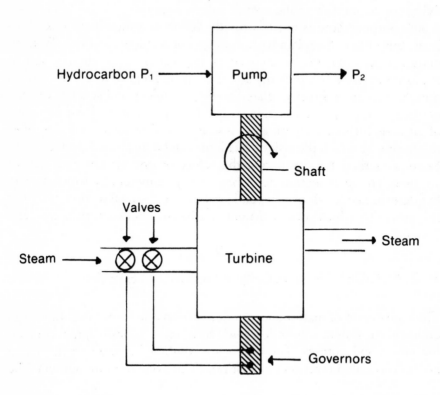

The subsystem described above is typical of the multitude of subsystems that perform a function in a manufacturing process, sporting event, or the home. These systems and components possess an *accident potential* or hazard that is related to (1) the mechanical, chemical, thermal or electrical energy that they store in normal operation, and (2) the energy that is released if they malfunction. For example, if the pipes leak, if the pump fractures, if the bearings are destroyed, then hydrocarbon liquid will be released into the atmosphere and will vaporize. The gas is combustible, and if it contacts oxygen in the presence of a spark, an explosion and fire may occur. If the fire, in turn, ignites a larger body of stored hydrocarbon, then even more energy is released, and more property damage will occur, along with possible personal injuries and/or fatalities. The *risk* associated with this system is directly related to the probability of the different malfunctions, and to the probable spectrum of consequences associated with each malfunction. The term "risk" therefore implies a cause-and-effect relationship between input events that could lead to energy releases, and output events (physical and psychological damage) that are the result of released energy.

The cause/effect aspect of an accident implies that any quantitative description of risk contains a measure of probability that a given failure event will occur (e.g., a fracture of the pump casing). This probability is determined by the *frequency, f,* of such incidents, which is the number of such failures that occur in a given time period (typically in one year or in one hour), divided by the number of units that are operating in the given time period. If 10,000 compressors of a particular class operate in one year, and one compressor typically fails per year, then the probability of compressor failure is 10^{-4} per year. The second part of the cause/effect relation, or the consequence, is the *severity, S,* associated with a given failure. The severity is related to the energy released by the failure event, ΔE, multiplied by the probability, p, that a sufficient portion of the energy impacts a person or object such that he/it is injured or malfunctions. If the second malfunction in turn leads to another energy release, which in turn leads to a third malfunction, then an *accident chain* has been created. In the case of the compressor/turbine system, a typical accident chain might be the following: vaporization of the liquid reduces drag on the turbine which causes the turbine shaft to rotate at a higher speed. In principle, the governor should sense this phenomenon and maintain steam flow rates such that the turbine does not overspeed. Suppose that the monitoring governor and the overspeed governor are both inoperative. Failure of the governors in turn allows turbine overspeed, which in turn leads to bearing failure in the compressor, which in turn allows air to enter the compressor. If the failure is detected at this point, the damage is minimal. However, if the shaft rotates eccentrically and at high speed for sufficient time that high frictional heat is generated in the compressor, the hydrocarbon/air mixture may explode, causing compressor rupture. The hydrocarbon then can spill out of the compressor at a high rate. If the system is shut down and fire prevention measures are used, the hazard can still be contained. However, there is a probability that the hydrocarbon will ignite, causing a large fire and extreme property damage and personal injury to workers. In this case, vaporization and governor malfunctions, due to poor maintenance or wear-out, have led to a spectrum of consequences. While the frequency

FIGURE 10.2

Automobile Accident Frequency-Severity Distribution, State of Texas, 1971

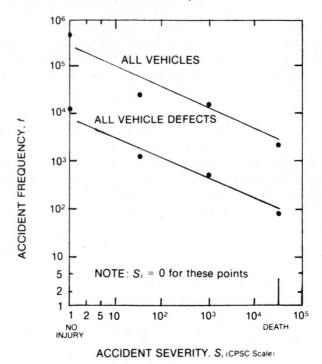

Source: A. S. Tetelman and M. L. Burack, "An Introduction to the Use of Risk Analysis Methodology in Accident Litigation" (6).

of one governor failure, f_1, may be 10^{-3} per year, the probability of fire due to governor failure is very low, perhaps as low as 10^{-15}, *because five separate 10^{-3} events must occur in order to produce the one, major 10^{-15} catastrophe.* This factor is the basis for one of the fundamental concepts of risk analysis—*that the more severe the accident, the less frequently it will occur* (cf. Figures 10.2, 10.3). This concept will be explored more fully in the following section, where some methods of quantifying risk are described.

FIGURE 10.3

FREQUENCY-SEVERITY FOR AUTOMOBILE ACCIDENTS
CAUSED BY VARIOUS DEFECTIVE VEHICLE SUBSYSTEMS AND COMPONENTS

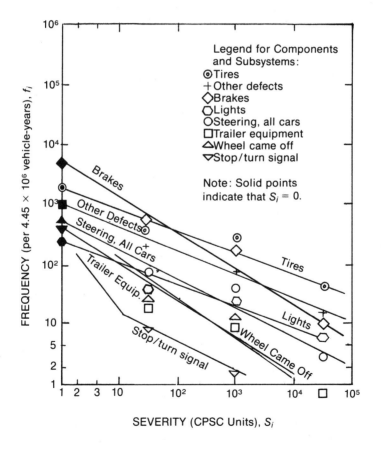

SOURCE: All data categories in Table 10.4 listing one or more fatalities.

10.3 THE QUANTIFICATION OF RISK

The fundamental notion that the risk associated with an event is a function of both the event's frequency and its severity is intuitively reasonable; that notion seems in keeping with the judgments we make as individuals, as well as the broader judgments society makes. This notion of risk also reflects modern concepts of reliability engineering. To the reliability engineer, there is no such thing as a perfectly safe product; for every product there is a non-zero probability of failure (small, perhaps, but still non-zero), and a non-zero probability that such a failure will result in injury to somebody (not severe, perhaps, but still an injury). In order to make a product safer, the engineer can decrease the frequency (i.e., the probability) of failure and of the resulting injuries, or he can decrease the severity of such injuries, or he can do both.

The simplest (in terms of data gathering) method of evaluating personal risk is to determine the number of fatalities associated with a given process or activity. While helpful in certain areas such as commercial aviation, the method is too simple to account for most accidents where a spectrum of consequences (e.g., minor to major injuries) can occur. In these cases, it is more realistic to consider an *average consequence,* and to evaluate risk in these terms.

The most direct method for evaluating the risk or hazard from a process or activity is to multiply the frequency of failures, f, by the *average severity, $\overline{S}$,* that results from the use or exposure of the process. This product, I, is called the *hazard index* or *risk index.*

$$I = f \times \overline{S} \tag{10.1}$$

The parameter $\overline{S}$ can be evaluated in terms of lost man days to society, dollar cost of accident, or outage time for equipment.

A detailed method for describing personal injury accidents during the use of consumer products has been developed by the U.S. Consumer Product Safety Commission (CPSC) (6) and serves as a good example of the quantification of risk and hazard. The CPSC operates the National Electronic Injury Surveillance System (NEISS) to estimate the frequency and severity of injuries incurred by individuals who are treated at 2 percent of the emergency rooms of U.S. hospitals. Injuries resulting from 800 consumer products are indexed, and the number of injuries from each product is multiplied by 50 to estimate the total number of injuries per year, f, in the U.S. population from that product. The CPSC also assigns an average severity value S_i for each of eight injury categories, as shown in Table 10.1. The severity value is related to the number of lost man days to society as a result of the accident. The average severity $\overline{S}$ of all injuries and fatalities due to a given product is obtained from the relation:

$$\overline{S} = \frac{\sum\limits_{i=0}^{i=8} f_i S_i}{\sum\limits_{i=0}^{i=8} f_i} \times \sum \overline{S}_i \qquad (10.2)$$

where f_i is the number of accidents that cause an injury of severity S_i, and f is the total number of accidents involving that product, including those "incidents" of zero severity (no injury). The CPSC quantifies "risk" in terms of a *hazard index, I,* by simply multiplying the accident frequency, f, by the mean severity, $\overline{S}$, for that product. Thus, from equation (10.1)

$$\text{Risk Index } I = f \times \overline{S}$$

and hazardous products are ranked in order of decreasing value of I. Table 10.2 lists the 33 most hazardous consumer products in the United States in 1974, as evaluated by the CPSC (6).

TABLE 10.1

CPSC-NEISS VALUES FOR ACCIDENT SEVERITY CATEGORIES (1974)

Severity Category	Representative Diagnosis	Severity Value
0	Incomplete or otherwise not acceptable data	0
1	Mild injuries/small areas, dermatitis, and sprains	10
2	Punctures, fractures	12
3	Contusions, scalds	17
4	Internal-organ injury	31
5	Concussions; cell and nerve damage	81
6	Amputations, crushing, and anoxia	340
7	All hospitalized category sixes	2,516
8	All deaths	34,721

TABLE 10.2

TOP 33 CPSC PRODUCTS NOT SELECTED BY PROPOSED METHOD (1)

Product Description	CPSC Rank	Product Description	CPSC Rank
Bicycles	1	Basketball	18
Stairs	2	Non-upholstered Chairs	19
Doors	3	Storage Furniture	20
Drain Cleaners	4	Unpowered Cutlery	21
Tables	5	Clothing	22
Beds	6	Paints	23
Football	7	Household chemicals	
Swings	8	(other than caustic)	24
Gasoline and kerosene	9	Money	25
Home Structures	10	Floors	26
Power Motors	11	Glass Bottles	27
Baseball	12	Washing machines	28
Nails	13	Matches	29
Bathtub and shower structures	14	Ladders	30
Gas space heaters	15	Sun lamps	31
Swimming pools	16	Home workshop saws	32
Gas ranges	17	Fences, non-electric	33

It should be noted that the CPSC system does not consider the time of exposure to or use of a given product. The frequency of injuries is computed on an annual basis for the United States, and thus represents *the risk to society as a whole.* The risk to an individual is different, since an individual will use a bed 2500 hours per year and a welding torch only 50 hours per year. Individual risk is therefore better quantified in terms of number of accidents per exposure hour, but this parameter is difficult to quantify since the usage spectrum of certain products (e.g., welding torches) varies from one individual to another.

The CPSC system can also be extended to quantify the risk due to individual subsystems in, say, an automobile (1, 5). In certain states, police officers are required to complete a form on each accident that involves measurable property damage or personal injury. Cause categories are included, along with consequences, and the data are stored in a form that enables parametric analysis to be made. When the various injury classifications are normalized to the CPSC classifications (shown in Table 10.1) the sources of automobile risk can be better defined.

Figure 10.2 indicates the inverse frequency/severity relationship for auto accidents that was mentioned earlier. It should be noted that the relation is the same, irrespective of whether the accident was caused by an apparent "defect" or resulted from driver or pedestrian error. Defective vehicles account for only 2 percent of the total risk of driving. Figure 10.3 shows the risk breakdown for the major subsystems of an automobile, along with the tabulated hazard indices. Again, the inverse frequency/severity relation applies for the different subsystems, although the slopes of the curves vary widely. The low slope for tires indicates that tire failures lead to higher severity accidents than, for example, failures of turn signals. This fact, coupled with the high tire-failure rate, combines to make tires almost as hazardous as all other automobile subsystems combined (Table 10.3).

TABLE 10.3

CALCULATED VALUES OF FREQUENCY, f,
MEAN SEVERITY, $\overline{S}$, AND CPSC-DEFINED TOTAL SEVERITY I
FOR SEVERAL VEHICLE SYSTEMS, MAJOR SUBSYSTEMS, AND COMPONENTS

	f	$\overline{S}$	$I = f\overline{S}$ *	% of driving risk	% of risk due to defective vehicles
All vehicles	6.6×10^5	197	1.3×10^8	100	
All defective vehicles	1.39×10^4	254	3.78×10^6	3.4	100
Tires	2455	695	1.7×10^6	1.6	45
Other defects	1341	509	6.82×10^5	0.6	18
Brakes	7065	92	6.5×10^5	0.6	17
Lights	410	552	2.27×10^5	0.2	6
All steering	587	280	1.64×10^5	0.14	4
Trailers	951	36	3.42×10^4	0.03	0.9
Wheels come off	585	47	2.75×10^4	0.02	0.7
Turn signals	428	32	1.37×10^4	0.01	0.3
Wipers	8	5	40	3×10^{-7}	10^{-5}
Pitman arm separations	8270	3.24	2.68×10^4	0.02	0.7

*All data have been normalized to the total vehicle usage, 4.45×10^6 vehicle-years, in the state of Texas in 1971.

10.4 THE MEANING OF ACCEPTABLE RISK

Having described the physical factors that contribute to the hazard associated with processes and activities, and having discussed some methods of quantifying risk, we are now able to address the question of what constitutes an "acceptable" risk. This question is basically at the core of the safety question. All products can probably be made "safer"; however, if the risk associated with their usage is "acceptable," there is an implication that the product is already safe enough.

Although the level of risk can be quantified in terms of frequency and severity as discussed earlier, the level of "acceptable" risk is a relative or comparative parameter, and varies from one individual to another and from one society to another. Certain individuals are more likely to undergo hazardous sporting activities, such as hang-gliding, than other individuals. The acceptable level of pollution, noise, and the speed limit on highways varies from one state to another. Consequently, acceptable risk is not an absolute parameter or engineering constant, but rather is a variable with respect to both geography and history. The United States has stronger occupational health and safety laws than non-western nations, and even in the United States, acceptable levels of risk to workers in critical occupations (e.g., miners) have decreased in the last century.

Since "acceptable risk" is a relative rather than invariant parameter, any discussion of acceptability is, by definition, comparative. Dr. Chauncy Starr, former dean of engineering at UCLA and currently president of the Electric Power Research Institute, has done the most original and extensive studies of what society has chosen as a level of acceptable risk. Much of the following material was developed by Starr, and the reader is referred to references 3 and 4 for in-depth treatment of the subject.

Starr has compared the probability of death from a variety of activities with variables such as natural risk, benefit, etc. His findings are summarized here. Briefly, the fatality rate in the United States from all causes (primarily old age and disease) is about 1 percent per year or about 10^{-4} percent per exposure hour (365 days $\times$ 24 hours). Hence, the natural probability of death is 10^{-6} per exposure hour. The probability of death from truly "pure chance" events such as lightning strikes is 10^{-9} per exposure hour. Figure 10.4 shows the risk of fatality due to motor vehicle accidents that occurred from 1900 to 1960. The graph indicates that when only a small fraction of the population was involved in the activity, the risk was high. As greater numbers of people became involved, the risk level decreased, and approaches an equilibrium value of 10^{-6}, the

FIGURE 10.4

RISK AND PARTICIPATION TRENDS FOR MOTOR VEHICLES

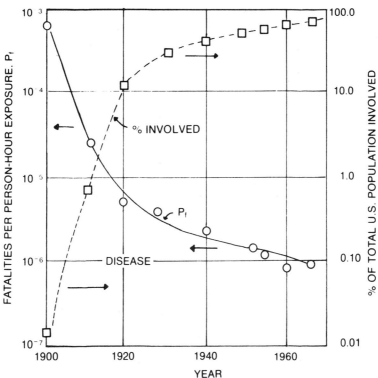

SOURCE: C. Starr, "General Philosophy of Risk-Benefit Analysis" (4).

risk level due to natural causes. Starr believes, and these views are shared by this author, that activities such as driving and flying (Figure 10.5) begin as sporting activities where one of the main attractions of the activity is the "macho trip" that is associated with risk levels in excess of the 10^{-6} norm. As time passes, individuals who enjoy the activity also try to find some profit as well as pleasure from it, and new industries form. Society becomes more involved and eventually more dependent on the activity, so that it loses its sportive character and becomes a natural activity which is an integral part of society. Consequently, through legislation and regulation, stricter safety codes and standards are imposed, and the risk level decreases until it reaches a natural level.

Many activities and technologies have developed out of sporting activities and games, as man has sought new pleasures, and out of fear, as man has sought to protect himself with improved medicine, housing,

FIGURE 10.5

RISK AND PARTICIPATION TRENDS FOR CERTIFIED AIR CARRIERS

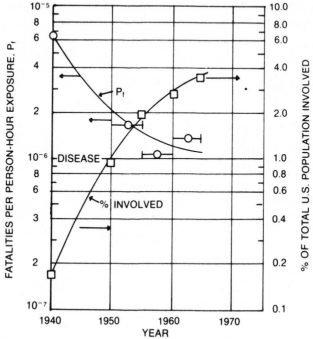

SOURCE: C. Starr, "General Philosophy of Risk-Benefit Analysis" (4).

and weaponry. Some activities, such as sports and transportation modes are *voluntary,* in that the consumer has a choice as to whether he wishes to participate and by which mode (e.g., driving vs. air travel). Other activities on which society spends its resources are non-voluntary. For example, the local utility company determines what type of fuel (coal, oil, nuclear) it will use in providing its customers with electric power, and the customers are *involuntarily* exposed to the risk associated with this power source. Figure 10.6 shows a plot of risk vs. benefit for a variety of activities, where benefit is measured in terms of dollars expended on the activity. Starr has found that the risk is proportional to the third power of benefit. He also notes that the risk level of voluntary activities is about one thousand times that of involuntary activities. In other words, individuals will accept higher risk levels, if they are free to choose them, than they will allow society to set for them.

These facts suggest some guidelines that society has subconsciously set for determining what constitutes acceptable risk. If the activity is voluntary, an acceptable risk level would be the natural level set by disease and old age. While driving and commercial aviation have reached this point, general aviation currently poses a risk that is about twenty-five times greater; this higher risk level will probably remain until more individuals become dependent on noncommercial aviation, and the activity becomes more integrated into our way of life. With respect to involuntary activities, it appears that society imposes a requirement that the activity be a thousand times safer. Consequently, in choosing to expose the population to an involuntary risk, some considerations need to be given to both the risk level and the benefits. Any activity whose risk is below that due to random ("Act of God") phenomena such as lightning strikes must be considered safe by any criteria, as society has no control over the probability that a random event will occur; society can only reduce this risk by minimizing the consequences of the event (e.g., through building codes designed to assure some earthquake resistance). The acceptable level of risk for other non-voluntary events should be related to the benefits of the event. As the benefit level increases, society should be willing to assume a higher risk level, although in no case should the level exceed that set by natural causes. The shaded area labeled "Involuntary" in Figure 10.6 shows the type of curve that would define the "acceptable risk" that is arbitrarily imposed on the individual.

FIGURE 10.6

RISK VERSUS BENEFIT: VOLUNTARY AND INVOLUNTARY EXPOSURE

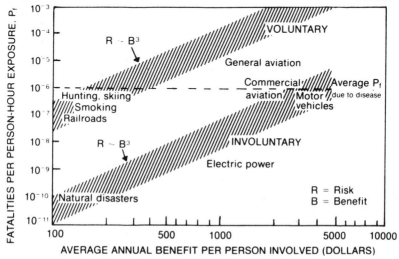

SOURCE: C. Starr, "General Philosophy of Risk-Benefit Analysis" (4).

While Starr's concepts have great merit in defining the scope of the "acceptable risk" question, systematic data dealing with naturally occurring levels of risk are often not available, and even when available, they may be either scanty or of questionable reliability. Moreover, the basic premise of the method—that the level of risk associated with such naturally occurring events as disease constitutes a threshold of acceptability somehow recognized (though perhaps tacitly) by society—is still as much a hypothesis as a proven fact. In fact, people often appear to find levels of risk associated with natural occurrences to be too high and to act to lower the levels of risk. Thus, people use lightning rods on their homes, and they have themselves and their families immunized against many common diseases.

There is, however, another method that can, in an important class of situations, be used to assess whether the risk associated with a particular kind of event is "acceptable." Suppose we are studying a system—an airplane or an automobile—that contains a particular part or component P, and we wish to determine whether the risk associated with a defect in part P is acceptable. Suppose further that the system has numerous parts and components besides part P, and that people customarily use comparable systems—comparable airplanes or automobiles—and accept the overall level of risk as a whole. Inherent in the overall level of risk is some degree of *natural variability*, reflecting the fact that accidents can never be prevented entirely, and that their frequency and consequences can never be predicted with certainty. Different severity values will be associated with different accidents resulting from a given product, and the frequency of accidents will not be perfectly uniform; thus, there will always be some *scatter* in the overall risk about an average value. *In such a situation, the level of risk associated with a defect in part P may be considered to be "acceptable" when that level of risk is not only smaller in magnitude than the average level of risk associated with the system as a whole, but is also as small as (or smaller than) the variations in the overall level of risk* (1, 5).

This method of determining what constitutes an "acceptable" risk seems intuitively reasonable. In voluntarily using the entire system, people unavoidably encounter and accept an average overall level of risk with an inherent variability. For instance, although there is an average level of risk associated with driving during the course of a year, the precise level of risk varies from time to time depending on many fctors, such as weather, road conditions, local population density, and whether it is a holiday weekend. If the risk associated with a defect in part P (a particular automotive component for instance) is about the same as or smaller than the unavoidable variations in the overall system risk, the risk associated with part P will look to the use of the system exactly like one of the unavoidable fluctuations in the overall risk. Since the user is willing to accept the overall risk together with its inherent variability, he should also be willing to accept the risk associated with the defect in part P. Table 10.3 shows an application of this method (5). In this particular case, a determination of the risk of automobile accidents due to failure of a steering component (the Pitman arm) was determined from analysis of failure frequency and severity. The data shown in Table 10.3 indicate that the risk due to Pitman arm failure is 0.02 percent of the risk of driving and 0.7 percent of the risk of driving a defective vehicle. As these percentages are well below the natural variations in risk, the particular component was considered to pose no unreasonable risk if continued in service.

10.5 METHODS OF RISK PREDICTION AND RISK REDUCTION

In the preceding section, we considered three primary criteria by which society, either consciously or subconsciously, determines whether a risk is acceptable:
1. Whether the risk is low compared to naturally occurring risk or to pure chance.
2. Whether the risk is justified in terms of benefit.
3. Whether the risk contributes a negligible amount to the total risk of an acceptable activity.

In making daily decisions about acceptability, all individuals, be they parents deciding whether to buy their son a skateboard, engineers deciding whether to add a second nondestructive inspection stage in a manufacturing process, or federal administrators determining whether to "recall" a product, face a major problem: *the data base available for use in their decision-making is minimal.* Most of the decisions are made on the basis of data bases (past history) ranging in sophistication from pure folklore, to the fact that two neighborhood children were injured on their skateboards, to the fact that a carefully selected and controlled sample of the population suffered undesirable side effects from a new drug.

The reliance on past history in making decisions about the future is, while imperfect, the only reasonable approach available. It would be impossible for an individual to calculate the risk based on evaluation of failure frequency, and all possible consequences of the failure. Indeed, this "event tree" approach is used only in large complex systems (e.g., nuclear power plants, space missions) where the acceptable failure rate is extremely small and the cost of these efforts is justified. In using past history, there are certain factors that need to be considered. The first factor deals with the nature of failure rates. Figure 10.7 illustrates the "bathtub" curve (so named because of its shape) that represents the rate of failure of a product, process, human being, etc. as a function of its age. When first introduced, unforeseen factors in design, errors in tooling and quality control, and lack of understanding of operators contribute to make the failure rate of a system or component relatively high. This period in the product's life is called the "infant mortality" period. After proper adjustments are made, the failure rate decreases to a constant level. During the "wear-in" period where the failure rate is low, constant, and "acceptable," those failures that do occur are the result of abusive loads that are placed on the system, overloads that are far in excess of what the system was designed to handle. The period where the failure rate finally increases is known as the "wear-out" or "end-of-life"

FIGURE 10.7

TYPICAL FAILURE RATE CURVE

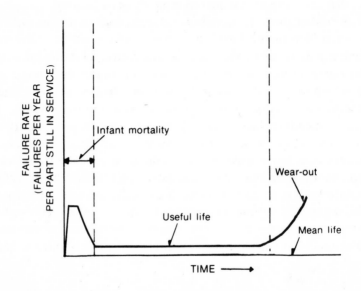

period. Here, failure occurs because of time-dependent degradation processes such as fatigue, corrosion, wear (for metals), and disease (for individuals) destroy the effectiveness of the components in the system. The mean life of the part, divided by some factor of safety, is then used to set an allowable operating lifetime.

If past experience or engineering analysis indicates that severe consequences could result from a failure, then a new product or process needs to be fully tested before it is placed on the market. The testing should be severe and in some cases the product should be proof tested at stress levels far in excess of operational levels, to determine its failure modes and its potential for failure due to abusive loading. Companies and others responsible for compliance with safety regulations need to observe the failure rate curve to determine whether a defect causing infant mortality remains in the system and whether the defect can be repaired or the component should be replaced. In large systems (e.g., autos, household appliances), evaluation of warranty complaints provides good indications of infant mortality, provided the data is collected and analyzed, but with smaller systems or individual components there is often no method for systematically evaluating the infant mortality period and for screening out defective parts prior to a major accident (which then attracts everyone's attention).

The acceptable failure rate during the wear-in period is related to risk level, as discussed earlier. From an engineering point of view, the real challenge is to make the proper separation between the design load and the abusive overload. Suppose that a gas pipeline is designed to operate at a certain pressure, P, without failing. Proper maintenance procedures can be set up to assure that cracks or corrosion do not form to lower the residual strength of the pipe. But suppose that a scraper or tractor impacts the pipeline, which produces a large crack, which leads to a large-scale brittle fracture, and then to an explosion from the escaping gas. Is this form of loading "reasonably foreseeable," and should it be accounted for in design? Should a manufacturer of bearings assume that proper lubrication schedules, although well outlined in company literature, will probably *not* be followed, and that there is a moderate probability that bearing seizure and overload fracture will occur? There are no well-defined answers to these questions, and, in the last analysis, the overall acceptability (or unacceptability) will be evaluated in terms of the total risk and its cost.

Even decisions relative to the wear-out period are complicated. If tests and analysis indicate that the mean life of a safety-related system on a light aircraft is 4000 hours, the manufacturer may suggest (and/or the FAA may require) that the system be replaced at 1000 hours, the safety factor of 4 being introduced to account for scatter in fatigue life. This approach to a safety problem often brings on complaints from aircraft operators that the part life should be *extended* to 2000 hours as they (the operators) have had no problems with the system. The inference is even made that a low allowable part life had been set so that the manufacturer could sell more spare parts. It is no wonder that manufacturers, accused on one hand of producing products with too high a failure rate, and on the other hand of replacing parts with too low a failure rate, often feel squeezed by competing interests of the government, environmental groups, and their customers.

In principle, and most often in practice, it is possible to monitor failure rates of major products and systems so that manufacturers, operators, and government officials can keep some control of failure frequency. The big unknown in risk assessment deals with the severity of consequences of a failure of a part, usually inexpensive, that is placed in the interior of some component. The failure of a fifty-cent accelerator linkage spring on an automobile can be handled by switching off the ignition and bringing the vehicle to a stop at the side of the road. But, if the failure occurs at a busy intersection, and the driver is too flustered to turn off the ignition, pedestrians may be injured or killed before the vehicle can be stopped.

Because of the inverse relation between failure frequency and severity (Figure 10.3) it is likely that hundreds of springs will fail before one fatal accident takes place. It is even possible to find and review spring failure data and determine that an accident of this type was probable. This review might suggest that the event was so foreseeable that the auto manufacturer should have installed a second, parallel spring into the system, so that one spring failure would not lead to system failure. The difficulty with this approach is that there are no simple methods available for predicting the slope of the frequency/severity curve for spring failure. In the absence of this information, it is not possible to make any quantitative analysis about the need for a second spring, and the engineer must rely on his best intuitive judgment.

In the absence of good engineering data on failure consequences, one can argue that it is better to err on the side of caution and when in doubt, make it safer. Federal regulations now make air bags optional on 1978 model autos but there is a drive to make the air bag mandatory, despite the lack of convincing data as to its effectiveness, initial cost, and maintenance cost. Many "judgment calls" such as these are not made on the basis of the state-of-the-art, but rather on the basis of society's attitude to "risk acceptance." Ten years ago, a mandatory air bag, while technically feasible, would have been considered an unnecessary item. Twenty years from now, society may be less preoccupied with safety than it is today and the air-bag concept may be obsolete.

The risk associated with a given product or process can be reduced by reducing the failure frequency, f, the average severity, $\overline{S}$, or by doing both of these things. Figure 10.8 shows the (f, S) graph divided into four regions. The upper right hand corner (Region 1) represents high frequency, high severity events, where the hazard is high and can only be reduced by drastic reductions in f, S, or both of these parameters. The lower left hand corner (Region 3) describes low frequency, low severity events that pose little hazard and are "safe."

In the other two areas (Regions 4 and 2) the risk is acceptable, as described by a curve of constant hazard I_a.

$$f = \frac{I_a}{S}$$

FIGURE 10.8

FREQUENCY-SEVERITY CURVE FOR ACCIDENTS

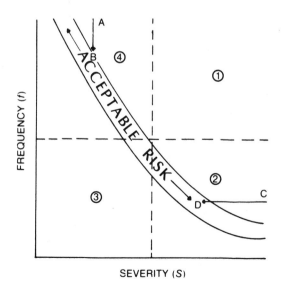

The curve is drawn as a band to account for the natural variation in risk, as described previously. In Region 4, most of the risk to society results from a high failure rate, F. These accidents generally involve individuals engaged in activities where small amounts of stored energy are released by a failure (e.g., a slip-and-fall, a burn from a cleaning solvent). In the absence of a protective covering for vulnerable parts of the body (e.g., helmets) it is unlikely that society can lower the risk by lowering the average severity, which is already low. Instead, improvements will come from lowering the failure rate (e.g., from A to B) through increased public awareness of the need for individual responsibility as it conducts activities such as going down stairs, riding bikes, etc. The majority of the most hazardous products listed by the CPSC (Table 10.2) fall into the high frequency, low severity category.

Region 2 deals with risk associated with low frequency, high severity events such as explosions, air crashes, and derailments. The far lower right portion of the region describes the real or perceived high severity event where no failures have occurred over a limited amount of service (e.g., catastrophic failure of a nuclear pressure vessel). In these instances, f is so ill-defined that part failure (and success) experience is not sufficient to make reliable predictions of risk. Risk assessments must then be based on the operational history of similar structures (e.g., non-nuclear pressure vessels). Primary reductions in risk in Region 2 can best be made through reductions in $\overline{S}$ (e.g., from C to D) through isolation of subsystems and structures that store large amounts of energy, such that failure in one subsystem does not trigger off a large accident chain.

10.6 SOCIETAL CONTROLS ON RISK

We have thus far considered risk in quasi-technical terms, without regard for the mechanisms that society has to control risk, and some discussion of this topic is in order. Traditionally, the marketplace provided the control on risk; if one product is safer than the other, and if safety is of concern to the individual, then he will choose the safer of two similar products when given the choice. Many risks, however, are involuntary; and even in voluntary activities such as hunting and skiing, the risk level associated with the purchasable hardware (e.g., rifles, ski bindings) is very low compared to the risk resulting from user error. The major voluntary decision that is available to the consumer is whether or not to purchase insurance, but this decision provides no protection and is important only if an accident occurs.

Decisions relative to risk reduction and accident prevention have, in the last decade, been made primarily by federal and state governments. Federal legislation has created a whole group of agencies, such as the Nuclear Regulatory Commission, the National Highway Traffic Safety Administration, and the Federal Aviation Administration, to name but a few. These agencies have the broad responsibility to determine whether drugs, automobiles, aircraft, and nuclear power plants are "safe." If these agencies decide otherwise, they have the authority to remove these products from service, and to require repair or upgrading (e.g., auto recall campaigns, grounding of a fleet, etc.). Manufacturers and operators who are faced with these recall decisions have the option of complying with them or of challenging these decisions in court. In effect, this has brought the judicial system directly into the decision-making process of what constitutes an acceptable risk.

The courts have traditionally been concerned with the question of reasonable risk on an after-the-fact basis, as accidents invariably lead to civil lawsuits between injured parties and/or corporations. Engineers often question whether lay judges and juries have the ability to render correct decisions in complex technical matters,* and corporations and insurance companies are concerned that a sympathetic jury will make awards that are grossly unfair.

This author has had the privilege of testifying as an expert witness in numerous product-liability litigations throughout the country. It is his opinion that the present system is both fair and satisfactory. Somehow, after all the conflicting evidence is presented, the juries are able to arrive at reasonable decisions, even if they have not understood all the complex technical facts. Similarly, in seven states where critics of nuclear power have sought to restrict its use severely through passage of initiatives, the voters have decided that the risk of a nuclear incident is more than justified by the benefit of having the necessary electric power, and have voted down the initiatives. These observations suggest that engineers and scientists should, like politicians, respect the ability of the public to determine its own best interest. †

The two major problems associated with risk assessment relate to "perceived" risk and to long-term effects of decisions made today. The major disaster gets major attention by news media and causes the public to consider the possibility of nuclear catastrophes, dangers from nuclear

---

*Refer to chapter 11 for current development in this regard.

† The reader is advised to refer to chapters 1 and 11 for further arguments on this.

sabotage, and leaks of radiation and other highly toxic substances. These perceived risks may be actually much lower than more mundane risks due to radiation damage from excess sunbathing. The perception issue is particularly acute when there is a possibility that not only the present generation, but future generations as well, will be influenced by regulatory decisions. This author takes the position that these decisions and risks are not different nor greater than those taken by individuals such as Queen Isabella and Columbus. The discovery of America led to profound changes in the life style of Europe, changes that could never have been allowed on the basis of the state of "risk acceptance" that existed in the sixteenth century. Likewise, it is not possible for engineers in the twentieth century to predict all possible forms of hazard and guard against them. Since a refinery manager cannot prevent an explosion by assuring that all overspeed governor devices are functional on his turbines, it is unreasonable to expect that engineers, legislators, and administrators can make meaningful, safety-related decisions that will have a guaranteed, non-destructive impact on future generations. The best method of assuring reasonable risk levels is to assure that capable, well trained individuals are involved in the decision-making and operation of safety-related systems, that proper data are kept and maintained, and that the public is made more aware of what risk is all about.

10.7 EXERCISES

1. Name two parts of the cause-effect relation of accidents. Can you relate these to an engineering device that you are familiar with?

2. Engineers make a product safer by decreasing the frequency of failure and/or decreasing the severity of resulting injuries. Can you cite any product for which this has been practiced?

3. What is the *hazard index* (also called *risk index*) and what is its purpose?

4. What is CPSC? and NEISS? What is their purpose?

5. Does the CPSC risk index system represent risk to the individual or to society as a whole? Explain.

6. Does driving a motor vehicle (*a*) today, (*b*) in the 1920s present more risk to the individual than purely natural causes? Discuss.

7. Study Figures 10.5 and 10.6 and write a short essay on the fatality rate and percentage of accidents versus technological improvements.

8. Why is the risk level of voluntary activities greater than that of involuntary activities? Would this mean that the technology is safer than the individual is willing to risk? Discuss the issue for specific cases.

9. Present at least two precautions people take to lower the level of risk associated with natural occurrences.

10. What is meant by *natural variability?*

11. What are the three primary criteria by which society determines whether a risk is acceptable? Give examples of each and discuss the interrelationship of these criteria for each case.

12. Name at least one technical device for each one of the four regions in Figure 10.8 and explain your reasons.

13. "Societal Controls on Risk." What does this mean? Give at least two examples and discuss their validity.

14. What is the role of the courts in regard to "acceptable risk"?

10.8 BIBLIOGRAPHY

1. Besuner, P. M., A. S. Tetelman, G. R. Egan, and C. A. Rau. "The Combined Use of Engineering and Reliability Analyses in Risk Assessment of Mechanical and Structural Systems." *Proceedings of Risk Benefit Methodology and Application Conference.* Asilomar Conference Grounds, December 1975, pp. 353–89 (UCLA ENG 7598).
2. *NEISS News.* National Electronic Injury Surveillance System. U.S. Consumer Product Safety Commission, 1973.
3. Starr, C. "Benefit-Cost Studies in Socio-Technical Systems." *Colloquium on Benefit-Risk Relationships for Decision-Making.* Washington, D.C., April 1971.
4. Starr, C. "General Philosophy of Risk-Benefit Analysis." Presented at EPRI/Stanford IES Seminar, Stanford, Calif., September 30, 1974.
5. Tetelman, A. S., and M. L. Burack. "An Introduction to the Use of Risk Analysis Methodology in Accident Litigation." *Journal of Air. Law and Commerce.* 1976, pp. 133–64. (Southern Methodist University School of Law, Dallas.)
6. United States Consumer Product Safety Commission, Computer Tabulated Accident Data Sheets. U.S. Consumer Product Safety Commission Bureau of Epidemiology, 1973–1974.

Chapter Eleven

CONTROLLING ENGINEERING DEMOCRATICALLY

11.1 TECHNOLOGICAL DISPUTES*

Two strongly contrasting doctrines can be set forth for the control of our very powerful technology. The first is the doctrine of the *moral responsibility of scientists and engineers.* This doctrine calls for technologists to take full responsibility for the consequences of technology. It holds that they should do their best to anticipate whether the knowledge they are discovering and the understanding of nature that they are creating will be used for good or for evil.

According to this doctrine, their duty is then to develop knowledge that they perceive to be good and to act to prevent the development of fields of knowledge that they believe will be harmful. It is argued that scientists and engineers who are engaged in opening up a new field have a deeper opportunity and more time to think about its moral and political consequences than does the general public. This superior opportunity demands that they act to prevent any harmful consequences they foresee.

*Extracted and edited from "Controlling Technology Democratically," by A. Kantro-. witz, *American Scientist.* September/October 1975, vol. 63, no. 5, pp. 505–9.

It goes without saying that the individual scientist-engineer has a personal responsibility for his own work. But his right to make personal moral choices about what he will work on is a separate issue from his responsibility to render scientific evaluations to the government or to society. This distinction must be most sharply drawn when the evaluation of scientific facts contributes to the making of large-scale decisions that control the development of technology.

In considering the consequences of governance of technology under the doctrine of the moral responsibility of scientists and engineers, it must first be recognized that much of our society already depends on what we do with advanced technology and still more will depend on our choice of which technologies to develop for the future. Second, the employment of the prestige of science to advance political and moral views raises very serious questions. It is, for example, unthinkable in a democratic society that scientists would actually be endowed with the authority to assume full moral responsibility for the social impact of science.

This view was expressed eloquently by Harold Laski in *The Limitations of the Expert* (9):

> It is one thing to urge the need for expert consultation at every stage in making policy; it is another thing, and a very different thing, to insist that the expert's judgment must be final. For special knowledge and the highly trained mind produce their own limitations which, in the realm of statesmanship, are of decisive importance. Expertise, it may be argued, sacrifices the insight of common sense to intensity of experience. It breeds an inability to accept new views from the very depth of its preoccupation with its own conclusions. It too often fails to see round its subject. It sees results out of perspective by making them the center of relevance to which all other results must be related. Too often, also, it lacks humility; and this breeds in its possessors a failure in proportion which makes them fail to see the obvious which is before their very noses. It has, also, a certain caste spirit about it, so that experts tend to neglect all evidence which does not come from those who belong to their own ranks. Above all, perhaps, and this most urgently where human problems are concerned, the expert fails to see that every judgment he makes not purely factual in nature brings with it a scheme of values which has no special validity about it. He tends to confuse the importance of his facts with the importance of what he proposes to do about them.

The implications of this doctrine of the moral responsibility of scientists have led Theodore Roszak to say in *The Making of a Counter Culture* (13):

> The key problem we have to deal with is the paternalism of expertise within a socioeconomic system which is so organized that it is inextricably beholden to expertise. And, moreover, to an expertise which has learned a thousand ways to manipulate our acquiescence with an imperceptible subtlety.

Roszak sees no solution to this key problem and concludes that technology is uncontrollable. He therefore advocates the return to what he calls "non-intellective thinking."

If fully applied, the doctrine of the moral responsibility of scientists-engineers leads to a kind of paternalistic control of society by a technological elite who will determine what is good for "the people." It amounts to a modern version of governance by *noblesse oblige.*

The second doctrine is one that we call *democratic control of technology.* Its essential feature is that decisions concerning which technology is good and which is evil are decided by the democratic process that gives each person one vote. It is difficult in today's America to retain any illusion that democratic process guarantees good government. However, we do not propose to discuss alternatives to democratic process, granting the wisdom of Churchill's aphorism that "democracy is the worst form of government, except all the others." Our purpose is simply to discuss the methodology of democratic control of technology. The essential problem is to find truth among the conflicting claims made by sophisticated advocates when there is serious controversy within the technological community.

How can the people or their elected representatives be helped to make informed decisions in the presence of such controversy? The need for a formal procedure was well illustrated by one of the debates in 1971 over whether or not to continue the development of the SST. In the last few weeks before the Senate vote, experts came forward with the claim that the operation of a fleet of SSTs would deplete the ozone in the upper atmosphere, allowing more ultraviolet radiation to reach the earth, which in turn would result in an increase in the incidence of skin cancer. This possibility was denied by equally competent experts, and 100 senators found themselves faced with the necessity for deciding their vote in part on the basis of an extremely complicated set of scientific claims that were being vigorously disputed among the experts. To the extent that their decision was swayed by this issue, no one in the Senate was really equipped to make a reasoned judgment.

Another example of a difficult decision involving new technology as well as value judgments is the question of how rapidly to reduce automobile pollution. Implementation of the Clean Air Act of 1970 called for reducing automobile emissions by about a factor of ten by 1976. The costs of this reduction using current technology have been estimated to be in the area of tens of billions of dollars per year, but many technical questions about feasibility and the health benefits that might result continue to give rise to controversy. Allan Mazur has collected "Disputes between Experts" (10) in which he clearly illustrates the need for a source of scientific judgment of higher presumptive validity.

The Mixed Decision

Decisions frequently faced by our government involving both technology and value judgments we will call mixed decisions. These decisions all involve extrapolation of known scientific fact or currently available technology and are of sufficient political or moral importance that divergences of opinion are bound to appear.

The essential input from the scientific community to decision-making in the United States is through the scientific advisory committee. Without going into detail about this process, we would like to make several points. In evaluating scientific advice on questions of great social importance, we must first recognize that the moral responsibility which many scientist-engineers feel very deeply can easily affect their judgment as to the state of scientific fact, especially when the pertinent scientific facts are not yet crystal clear. Second, it should be noted that the selection of scientific committees has always been beset by the dilemma that one must choose between those who have gone deeply into the subjects under discussion and who, accordingly, have preconceived ideas about what the outcome should be, and those who are perhaps unprejudiced but also relatively uninformed on the subjects under discussion.

Finally, the fact that scientific advisory committees have, in many cases, played an influential role in decision-making without taking public responsibility for their judgments warrants serious concern. In the making of mixed decisions, the validity of the scientific input has frequently been brought under question.

It has occasionally been maintained that the scientific and non-scientific components of a mixed decision are generally inseparable. It is, of course, true that a final political decision cannot be separated from scientific information on which it must be based. The reverse is not true; a scientific question which, logically, can be phrased as anticipating the results of an experiment can always be separated from any political considerations. Thus, the question—Should we build a hydrogen bomb?—is not a purely scientific question. A related scientific question—Can we build a hydrogen bomb?—could in principle be answered by an experiment.

It is almost inevitable that scientists who have been engaged in research relevant to the scientific side of great mixed decisions should have deeply held political and moral positions on the relationship of their work to society. Scientific objectivity, a precious component of wise mixed decisions, is thus very difficult to achieve. We do not believe it is possible for scientists to have deeply held moral and political views about a question and simultaneously maintain complete objectivity concerning its scientific components.

In the past, moreover, scientific advisory committees have frequently developed close relationships with the officials who have final decisions to make. They have frequently advised political figures about what final decisions they should reach, not only about the scientific components of a decision but about the moral and political implications as well. Although the close relationship may be valuable, it does point up a need for an alternative source of scientific judgment that will forgo taking any moral or political stands and will seek to optimize objectivity.

The separation of scientific from nonscientific components of a mixed decision is the key proposition we have to make. It is the old issue of the separation of facts from values, and we submit that this separation can always be made. In order to maintain democratic control of mixed decisions, it is essential that great care be taken to avoid the invasion of objectivity by strongly held moral or political views.

11.2 THE SCIENCE COURT*

There are many cases in which technical experts disagree on scientific facts that are relevant to important public decisions. Nuclear power, disturbances to the ozone layer, and food additives are recent examples of these mixed decisions. As a result, there is a pressing need to find better methods for resolving factual disputes to provide a sounder basis for public decisions. The Task Force accordingly proposes a series of

---

*Based on "The Science Court Experiment: An Interim Report," by the Task Force appointed by the Presidential Advisory Group on Anticipating Advances in Science and Technology, *Science*. August 20, 1976.

The task force is composed of three members of the presidential advisory group—

experiments to develop adversary proceedings and test their value in resolving technical disputes over questions of scientific fact.* One such approach is embodied in a proposed Science Court that is to be concerned solely with questions of scientific fact. It will leave social value questions—the ultimate policy decisions—to the normal decision-making apparatus of our society, namely, the executive, legislative, and judicial branches of government as well as popular referenda.

In many of the technical controversies that are conducted in public, technical claims are made but not challenged or answered directly. Instead, the opponents make other technical claims, and the escalating process generates enormous confusion in the minds of the public. One purpose of the Science Court is to create a situation in which the adversaries direct their best arguments at each other and at a panel of sophisticated scientific judges rather than at the general public. The disputants themselves are in the best position to display the strengths of their own views and to probe the weak points of opposing positions. In turn, scientifically sophisticated outsiders are best able to juxtapose the opposing arguments, determine whether there are genuine or only apparent disagreements, and suggest further studies which may resolve the differences.

The Task Force has no illusions that this procedure will arrive at the "truth," which is elusive and tends to change from year to year. But we do expect to be able to describe the current state of technical knowledge and to obtain statements founded on that knowledge, which will provide defensible, credible, technical bases for urgent policy decisions.

Dr. Arthur Kantrowitz, AVCO Everett Research Lab, Inc. (chairman); Dr. Donald Kennedy, Stanford University; and Dr. Fred Seitz, Rockefeller University—and the Honorable Betsy Ancker-Johnson, U.S. Department of Commerce; Mr. David Beckler, National Academy of Sciences; Dr. Edward Burger, Georgetown University Medical Center: Mr. William Cavanaugh, ASTM; Dr. Russell C. Drew, National Science Foundation; Mr. William Holt, U.S. Department of Commerce; Dr. Paul Horwitz, Congressional Fellow; Honorable Lawrence Kushner, Consumer Products Safety Commission; Professor Allan Mazur, Syracuse University: Dr. Joel Primack, University of California, Santa Cruz; Mr. Sheldon W. Samuels, AFL-CIO; Honorable Richard O. Simpson, Consumer Products Safety Commission; Mr. Donal Strauss, American Arbitration Association; Mr. David Swankin, Swankin and Turner; Dr. Myron Tribus, Massachusetts Institute of Technology; and Mr. James S. Turner, Swankin and Turner.

*We use the expression "scientific fact" to mean a result, or more frequently the anticipated result, of an experiment or an observation of nature.

The basic mechanism proposed here is an adversary hearing, open to the public, governed by a disinterested referee, in which expert proponents of the opposing scientific positions argue their cases before a panel of scientist/judges. The judges themselves will be established experts in areas adjacent to the dispute. They will not be drawn from researchers working in the area of dispute, nor will they include anyone with an organizational affiliation or personal bias that would clearly predispose him or her toward one side or the other. After the evidence has been presented, questioned, and defended, the panel of judges will prepare a report on the dispute, noting points on which the advocates agree and reaching judgments on disputed statements of fact. They may also suggest specific research projects to clarify points that remain unsettled.

The Science Court is directed at reducing the extension of authority beyond competence, which was Pascal's definition of tyranny. It will stand in opposition to efforts to impose the value systems of scientific advisors on other people. As previously stated, the Science Court will be strictly limited to providing the best available judgments about matters of scientific fact. It is so constructed in the belief that more broadly based institutions should apply societal values and develop public policies in the areas to which the facts are relevant.

Procedures

Issue Selection. The word issue is used here to refer to a decision pending before a governmental agency. These decisions will frequently be mixed ones involving important social values as well as controversial scientific facts. Described below is a procedure through which questions of scientific fact can be separated from value-laden issues. Some examples of issues under consideration are: Should fluorocarbons be banned because of their impact on the ozone layer? Is Red Dye #40 safer than Red Dye #2? Should water supplies be fluoridated? We do not presently intend to use the nuclear power issue as a subject in the initial experimets with the Science Court concept. Later it is hoped that a developed Science Court will be able to contribute to the making of public policy even on as divisive and pervasive an issue as nuclear power. Issues to be examined in the experiments will be selected according to three criteria:

1. Issues must be policy-relevant and have technical components that are both important and apparently disputed.
2. Issues allowing easy separability of facts from values will be favored for the experiments.

3. Issues will be favored for which informed and credible case managers (adversaries) can be obtained. To simplify the process, it will be valuable to choose an issue in which two case managers can fairly represent all facets of the controversy.

Funding. Frequently the opposing parties to a technical controversy have vastly different resources available to them. The Task Force sees no way to eliminate such inequalities but it is certainly imperative that each side be provided with sufficient funding to prepare an adequate presentation for the Science Court.

Considerable doubt has been expressed about the wisdom of seeking funding directly by a government agency involved in the issue. It is argued that, although money could be given without strings, there might be an implication that the next time the Science Court came for funds the agency's decisions would depend on whether the first ruling was "acceptable." Therefore, it has been suggested that initial funding come from the National Science Foundation (NSF). In addition to the NSF, there would be considerable advantage in having a variety of funding sources for the Science Court experiment, including private foundations or business sources. In every case assurances must be had that no strings are attached.

It is important to have involvement of an agency in whose jurisdiction the issue falls so that it can help in formulating the issue, advise on the procedure, and provide necessary power to compel release of relevant information.

Selection of Advocates. Once an issue has been selected, and funding obtained, the next step is to choose the adversaries, specifically a chief adversary for each side, to be called the "case managers." Two procedures are currently under consideration.

1. The Science Court or a collaborating agency issues Requests for Proposals (RFPs) for case managers. Each submitted proposal should exhibit that the bidder has the expertise and constituency to speak for one side of the issue and name its case manager. For example, a group such as the Union of Concerned Scientists, the Sierra Club, or Friends of the Earth might be a reasonable bidder to represent the anti–nuclear power side of that issue. It might form an alliance with a scientific institution such as a nonprofit analysis group and/or individual consultants. In any case, the objective is to exhibit that the bidder can provide the best case for its side of the issue. Combinations of groups opposing

nuclear energy would be encouraged, and the RFP would point out that such coalitions will be favored to receive the contract. In this example, the Atomic Industrial Forum might well bid to represent the side favoring nuclear energy though conceivably it would choose to join other scientific groups.

The scientific credentials and constituency of the proposers will be examined carefully by the Science Court and/or the collaborating agency, and a selection made by processes similar to those used in selecting contractors for other purposes. The two chosen case managers will then be funded to participate in the procedure outlined below, perhaps on a time and materials basis or by some other suitable contractual mechanism.

2. When an issue is clearly polarized, the case managers might be found by polling the interest groups involved on each side.

Selection of Judges and Referees. It is currently envisioned that the Science Court with consultation from appropriate scientific societies and organizations will produce a list of prospective judges certified as unusually capable scientists having no obvious connections to the disputed issue. They will then be examined by the case managers for prejudice. After acceptance, a panel of judges, say, three for the first experiment, will be formed.

In addition to the panel of judges, there should be a Science Court–selected referee who is concerned with the implementation of agreed procedures in a scientific setting. For discussion it is proposed that the referee should be a scientist advised by legal counsel, so that full responsibility for this procedure can be retained by the scientific community.

Several questions are still under discussion concerning these functions. One is whether the role of referee should be undertaken perhaps by a chief judge advised by legal counsel. This might simplify the organizational structure and centralize the authority necessary to maintain an orderly procedure. Another question has been raised as to whether the prospective judges should be selected by "elite" institutions such as the National Academy. It might be advantageous to have some prospective judges chosen by random selection from competent members of the various professional societies.

Transition, Issue to Factual Questions. As was pointed out previously, an issue selected for a Science Court experiment will be an issue that is before a government agency. It is most important that the issue be stated in a manner as close as possible to the actual decision which must be made by the agency. Thus, the Task Force proposes to prevent selections of a part of the issue that might prejudice the result. For example, the issue would not be: "Are nuclear power plants explosive in the sense of an atomic bomb?" but, "Should a specific nuclear plant be licensed or not licensed?" The broader question will provide the case managers with an opportunity to state all of the scientific facts that they consider important to their case. Selecting the narrower issue concerning explosive potentialities would be prejudicial because a negative answer (conceded we believe by most participants in this dispute) would be prejudicial without affording case managers a full opportunity to develop the facts basic to their opinions.

The selected issue will probably be a value-laden controversial matter. It is proposed that the Science Court go through a process by which factual questions under dispute can be isolated. The first step is the formulation by the case managers of a series of factual statements which they regard as most important to their cases. Factual statements must conform to the definition given earlier—they must be results of experiments or observations of nature. This definition excludes statements such as "if X occurs, then Y *may* occur." Such a statement is valid even if the probability of the occurrence of Y is infinitesimally small, so the experiment required to refute the statement is impossible. An acceptable version of the statement must specify a finite probability which could be refuted by a possible experiment.

After the statements have been examined by the referee or the judges to be sure they are confined to statements of scientific fact, the statements will be exchanged between case managers. Each side is then invited to accept or to challenge each of the opposition's statements. Since the statements are drafted in the knowledge that they will be subjected to sophisticated challenge, it is hoped that exaggeration and vague language will be deemed counterproductive. Therefore, many or even all of the statements made may not be challenged. In this case, the Science Court procedure will have been extremely successful in coming forth with an accepted series of factual statements.

Challenges. The case managers will examine the lists of statements of fact made by their opponents and decide which they can accept and which they challenge. The challenged statement will first be dealt with by a mediation procedure in which attempts are made to narrow the area of disagreement or to negotiate a revised statement of fact that both case managers can accept. If this procedure does not result in an agreed-upon statement, the challenge will be the subject of an adversary procedure.

Adversary Procedures. Several important aspects of the adversary procedure are still being worked out. First, it must be decided to what extent the experimental Science Court will be able to compel disclosure (employing legal powers vested in the collaborating government agency) of scientific information by such processes as subpoena, discovery, etc. A second important matter under discussion is the relative desirability of keeping the rules of procedure flexible enough to allow a more rapid development of fair and effective procedures versus the probable necessity of fixing the rules before the case managers agree to accept the Science Court procedure. We propose now to have the initial rules agreed upon by the case managers and changed only with the agreement of both case managers during the experiment or at the start of a new experiment.

The adversary proceeding will begin with a case manager putting forth his substantiation of a challenged statement in the form of experimental data and theoretical calculations. This evidence will be subjected to detailed scrutiny conducted in the tradition of a scientific meeting but with the added discipline of adhering closely to the challenged statement. It is important to recognize that the applied rules of evidence will be the scientific rules of evidence and not the legal rules of evidence. Thus, *ad hominem* attacks will be ruled out. There will be no necessity to prove the expertise of a witness since his statements are open to detailed challenge. We are unaware of any codification of the rules of scientific evidence, and intend to proceed at the outset on the simple statement that we will observe the rules that are traditional in the scientific community. On the other hand, we have a great deal to learn from the legal community on procedures. For example, the Science Court should not proceed unless representatives of both case managers are present. It should preserve the right of each case manager to cross-examine completely the positions taken by his adversary.

Considerable discussion has taken place regarding the degree to which the challenge-resolution procedure should be conducted in writing or orally. The advantages of a written procedure are:

1. It might make it easier to guard against dramatic oral presentations obscuring the merits of a case.
2. It might be easier to avoid the difficulties of "heavy" legal procedures.

3. It might well be more acceptable to the scientific community and more consistent with its traditions.

On the other hand, some members of the Task Force insist that an important part of the procedures should be oral. The advantages are:

1. The process could proceed more rapidly.
2. An oral presentation makes public observation and public scrutiny easy, and this is essential for credibility.

The complete proceedings of the Science Court will be open to the public with special provisions for the protection of proprietary information when necessary. However, the judges' deliberations after hearing the evidence should be conducted in private as in legal procedure.

An initial trial procedure is being drafted. However, the Science Court should not be bound by precedents but should continuously seek to refine its procedures to produce factual statements of the highest presumptive validity consistent with time constraints.

Results of the Proceeding

The primary results to be expected are a series of factual statements which will be arrived at in two ways. First, there will be the statements of fact made by the case managers and not challenged by their opponents. A second group of results will be the opinions of the judges regarding statements that were challenged. Some or most of these statements of fact will be qualified with statements about probable validity or margins of error. An important secondary consequence will be the lines drawn between areas where scientific knowledge exists and where it does not exist. Since important knowledge that is lacking will be pointed out, judgments of the Science Court will suggest areas where new research should be stimulated. In almost all cases the boundary between knowledge and ignorance will continuously shift, and revisions to take account of new knowledge may have to be made frequently when issues of great national importance are at stake.

It bears repeating that the Science Court will stop at a statement of the facts and will not make value-laden recommendations.

Evaluation of the Experiment

Any attempt to evaluate the outcome of this experimental adversary procedure is susceptible to bias. A prime entry point for bias is the initial decision of what it is about the project that will be evaluated. If it was decided to examine only those features of the adversary process that seem, a priori, trouble-free, then the evaluation is likely to come out positive; conversely, if attention is limited to troublesome features of the process, then the overall evaluation will almost certainly come out more negative. Therefore, it is essential to examine all those aspects of the experiment that are crucial to an informed decision on whether or not it "worked."

It seems useful to evaluate the *operation* of the Science Court separately from the *effect* of the judges' decision. By "operation" we mean the behavior of the Science Court's principals—case managers, judges, and referee. By "effect" we mean the alteration (if any) of attitudes and behavior of people outside of the experiment—regulatory agencies, industry, the mass media, legislators, interested citizen groups, and the wider public.

Operation. At a minimum, we need to know if the various principals fulfilled their assigned roles. Did they stick to questions of fact, avoiding value issues? Did the case managers agree on the selection of judges? Did they perceive themselves, and were they perceived by the other principals, as having made credible cases for their sides? Was the referee successful in keeping the other principals to the codified procedures? Were the codified procedures themselves satisfactory? Did the principals perceive that the judges reached reasonable and unbiased conclusions?

The evaluation should be as objective as possible, but we must recognize the great potential for a biased selection of small bits of data from the volume of experimental data, and also for a biased interpretation of data. Perhaps it would be useful to use three evaluators: one intending to present objective conclusions, one whose intent is to provide a positive picture of the experimental result, and one whose intent is to provide a critical picture. Ultimate evaluation of the experiment will benefit from exposure to these three diverse viewpoints.

Effect. At a minimum, we need to know if partisans perceive that "their" case manager did a credible job in making the case. Do they consider the procedures of the Science Court to be fair, even if they feel that their side "lost"? Do partisans change any of their attitudes or behavior as a result of the Science Court findings? Do regulatory agencies or other relevant governmental bodies take actions that appear to be based on the findings? Do they take contrary actions? Do the mass media provide accurate coverage of the debate and do they accept the findings? Are members of the wider public aware of the experiment? If so, do they understand the procedure, and do they know the Science Court findings? If so, do they express opinions that are consistent with the findings, even when they held contrary views prior to the hearing?

The answer to these questions can only be attained by creation of the proposed Science Court as an institution for scientific judgment. The future of such an institution would depend on the degree to which political and scientific communities would accept its initial judgments in comparison with the judgments reached by existing procedures. It seems possible that with a relatively modest start, an institution could be developed which, in the course of time, would achieve a much higher level of presumptive validity than now exists in communication between the scientific community and Congress. Such an institution could be invaluable in providing an improved scientific basis for future mixed decision of the Congress.

11.3 EXERCISES

1. Can you name some important technological issues that need control? Why?

2. What kind of control mechanism do we presently have for technological developments? Write a paper on the subject.

3. How do you interpret "noblesse oblige"? Write a short essay on the subject.

4. Name a dozen technological issues that can be brought to the Science Court.

5. How would you like to see the judges, advocates, and referees selected for the Science Court?

6. How would you interpret the "impossibility of separation of facts from values"? Can you argue this in a short essay?

7. Would you perceive any conflict between our federal judiciary system and the proposed Science Court?

11.4 BIBLIOGRAPHY

1. Boffey, Philip M. "Science Court: High Officials Back Test of Controversial Concept." *Science.* vol. 194, October 8, 1976, pp. 167–69.
2. Casper, Barry M. "Technology Policy and Democracy: Is the Proposed Science Court What We Need?" *Science.* vol. 194, October 1, 1976, pp. 29–35.
3. Cavanaugh, William T. "The ANSI Councils: A Democratic Route to Technology Assessment." New York: American National Standards Institute, February 1976.
4. Carroll, James D. "Participatory Technology." *Science.* vol. 171, February 19, 1971, pp. 647–63.
5. Hammond, Kenneth R., and Leonard Adelman, "Science, Values and Human Judgment." *Science.* vol. 194, October 22, 1976, pp. 389–95.
6. Kantrowitz, Arthur. "Controlling Technology Democratically." *American Scientist.* vol. 63, no. 5, September/October 1975, pp. 505–9.
7. Kantrowitz, Arthur. "Proposal for an Institution for Scientific Judgement." *Science.* vol. 156, no. 3776, May 12, 1967, pp. 763–64.
8. Kantrowitz, Arthur. Testimony before the Subcommittee on Government Research of the Committee on Government Operations, United States Senate, Ninetieth Congress, March 16, 1967. *Congressional Record.* June 8, 1967, p. 15256.
9. Laski, Harold. "The Limitations of the Expert." Fabian Tract No. 235. London: The Fabian Society, 1931.
10. Mazur, Allan. "Disputes between Experts." *Minerva.* vol. 11, no. 2, April 1973.
11. Mitroff, Ian I., and John A. Nelson. "An Experiment in Dialectic Information Systems." *Journal of the American Society for Information Sciences.* vol. 25, no. 4, July/August 1974.
12. Mitroff, Ian I., and Murray Turoff. "The Whys behind the Hows." *IEEE Spectrum.* vol. 10, no. 3, March 1973, pp. 62–71.
13. Roszak, Theodore. *The Making of a Counter-Culture.* New York: Doubleday, 1969.
14. Tribus, Myron. "A Proposal for Shaping Technical Information in the Public Interest." Letter to the Editor, *Astronautics & Aeronautics.* February 1972.
15. Wolf, Robert L. "Trial by Jury: A New Evaluation Method." *Phi Delta Kappan.* November 1975, pp. 185–87.

INDEX

303.483
K 18

112 340

DATE DUE